The Europeanization of British Environmental Policy

One Europe or Several?

Series Editor: **Helen Wallace**

The **One Europe or Several?** series examines contemporary processes of political, security, economic, social and cultural change across the European continent, as well as issues of convergence/divergence and prospects for integration and fragmentation. Many of the books in the series are cross-country comparisons; others evaluate the European institutions, in particular the European Union and NATO, in the context of eastern enlargement.

Titles include

Andrew Cottey, Timothy Edmunds and Anthony Forster (*editors*)
DEMOCRATIC CONTROL OF THE MILITARY IN POSTCOMMUNIST EUROPE
Guarding the Guards

Andrew Jordan
THE EUROPEANIZATION OF BRITISH ENVIRONMENTAL POLICY
A Departmental Perspective

Helen Wallace (*editor*)
INTERLOCKING DIMENSIONS OF EUROPEAN INTEGRATION

One Europe or Several?
Series Standing Order ISBN 0-333-94630-8
(*outside North America only*)

You can receive future titles in this series as they are published by placing a standing order. Please contact your bookseller or, in case of difficulty, write to us at the address below with your name and address, the title of the series and the ISBN quoted above.

Customer Services Department, Macmillan Distribution Ltd, Houndmills, Basingstoke, Hampshire RG21 6XS, England

The Europeanization of British Environmental Policy

A Departmental Perspective

Andrew Jordan

Lecturer in Environmental Politics, University of East Anglia
Norwich, UK

First published 2002 by
PALGRAVE MACMILLAN
Houndmills, Basingstoke, Hampshire RG21 6XS
and 175 Fifth Avenue, New York, N.Y. 10010
Companies and representatives throughout the world

PALGRAVE MACMILLAN is the global academic imprint of the Palgrave Macmillan division of St. Martin's Press, LLC and of Palgrave Macmillan Ltd. Macmillan® is a registered trademark in the United States, United Kingdom and other countries. Palgrave is a registered trademark in the European Union and other countries.

ISBN 0–333–94631–6

This book is printed on paper suitable for recycling and made from fully managed and sustained forest sources.

A catalogue record for this book is available from the British Library.

Library of Congress Cataloging-in-Publication Data
Jordan, Andrew, 1968-
    The Europeanization of British environmental policy: a departmental perspective/Andrew Jordan.
      p. cm. – (One Europe or several?)
    Includes bibliographical references and index.
    ISBN 0-333-94631-6 (cloth)
      1. Environmental policy–Great Britain. 2. Environmental policy–European Union countries. I. Title. II. Series.
GE190.G7 J67 2002
363.7'056'0941 – dc21   2001050802

10 9 8 7 6 5 4 3 2 1
11 10 09 08 07 06 05 04 03 02

Printed and bound in Great Britain by
Antony Rowe Ltd, Chippenham and Eastbourne

To Susan

Contents

List of Tables and Figures

Tables

Figures

Acknowledgements

This book has its origins in an MSc thesis on the implementation of European Union (EU) environment policy, which I completed in 1990. Back then, I assumed that the most important levers of British policy resided in Whitehall – or, to be more specific, the three-towered block of offices on Marsham Street, London, which (until the mid-1990s) housed the national Department of the Environment (DoE). The more I studied contemporary British environmental policy, the more evidence I uncovered of a deep-seated but greatly overlooked shift in the locus of British environmental politics and policy, arising from the continuing transfer of legal powers from Whitehall to Brussels. Over the years, I looked in vain for an account of the increasingly dense networks of daily interaction and communication that run between the main Whitehall departments and the institutions of the EU. I could not even find a comprehensive history of the Department, let alone a scholarly analysis of how it fitted into the structures and processes of British environmental policy-making. On my travels I found countless books on environmental pressure groups, environmental ideologies and environmental law, but virtually nothing on what is arguably *the* single most important environmental actor in Britain, namely the national environmental department, and its evolving relationship with the single most important driver of contemporary British environmental policy, namely the European Union.

This book is my attempt to fill that gap in current scholarship, although it is deliberately neither a detailed history of the Department (since 2001 the Department for Environment, Food and Rural Affairs, or DEFRA) nor a comprehensive account of how the EU has affected British environmental policy. Rather, it tries to assess the DoE's role in mediating the political frictions that have arisen since 1970 as British and EU environmental policy slowly entered one another's political orbit.

I am indebted to the following people for discussing the ideas presented in the following pages: Tim O'Riordan, David Lewis, John Rowcliffe, Tony Fairclough, Fiona McConnell, Derek Osborn, David Fisk, Nigel Haigh, Martin Smith, Claudio Radaelli and Albert Weale. Simon Bulmer, Helen Wallace, Rüdiger Wurzel, Neil Carter and Jenny Fairbrass commented on the entire manuscript in draft and saved me from innumerable errors. John Stevens of the Department of Transport, Local Government and the

Regions (DTLR) kindly granted me privileged access to a number of internal DoE documents, commented on various draft chapters and invited me to present my findings to a seminar held in the Department. Jenny (who co-authored Chapter 8 on biodiversity policy) and I would also to express our sincere thanks to John and all the other government officials, politicians and pressure group campaigners who agreed to talk about their experiences of 'the DoE and Europe'.

The research reported in this book was generously funded by the UK Economic and Social Research Council (ESRC) under grant number R000237870. I am especially grateful to Helen Wallace for inviting me to participate in the ESRC's 'One Europe or Several?' research programme (1999–2003) and enthusiastically promoting my work at every available opportunity. I would also like to thank Nigel Haigh for allowing me to use the Institute of European Environmental Policy's extensive 'library' of material in London. The staff of the British Library in London and the DoE/Department for the Environment, Transport and the Regions' own internal library also helped me to track down a number of important official publications that are not widely known or consulted.

Last but not least, I would like to thank Susan for her enduring love and friendship.

ANDREW JORDAN

List of Abbreviations

AQS	Air Quality Standard
BAT	Best Available Technology/Techniques
BATNEEC	Best Available Technology/Techniques Not Entailing Excessive Cost
BPM	Best Practicable Means
CBI	Confederation of British Industry
CDEP	Central Directorate for Environmental Protection (DoE)
CEC	Commission of the European Communities
CLA	Country Landowners' Association
CoM	Council of Ministers
COREPER	Committee of Permanent Representatives
CPRE	Council for the Protection of Rural England
CUEP	Central Unit for Environmental Pollution (DoE)
DEFRA	Department for Environment, Food and Rural Affairs (2001–)
DETR	Department for the Environment, Transport and the Regions (1997–2001)
DG-Environment	Directorate-General of the Environment (European Commission, formerly DG XI)
DoE	Department of the Environment (1970–97)
DoT	Department of Transport (1976–97)
DTI	Department of Trade and Industry
EC	European Community
ECJ	European Court of Justice
ECPS	Environment and Consumer Protection Service (European Commission)
EEB	European Environmental Bureau
EEC	European Economic Community
EIA	Environmental Impact Assessment
ENDS	Environmental Data Services Ltd
EP	European Parliament
EPEUR	Environmental Protection Europe Division
EPG	Environmental Protection Group
EPI	Environmental Policy Integration
EPINT	Environmental Protection International Division
ESRC	Economic and Social Research Council
EU	European Union

EUREAU	Union of European Associations of Water Suppliers
FCO	Foreign and Commonwealth Office
FoE	Friends of the Earth
HI	Historical Institutionalism
HMIP	Her Majesty's Inspectorate of Pollution
HOCESCA	House of Commons European Standing Committee (A)
HOLSCEC	House of Lords Select Committee on the European Communities
IAPI	Industrial Air Pollution Inspectorate
IEEP	Institute of European Environmental Policy
IGC	Inter-governmental Conference
IPC	Integrated Pollution Control
IPPC	Integrated Pollution Prevention and Control
LI	Liberal Inter-governmentalism
MAFF	Ministry of Agriculture, Fisheries and Food
MEP	Member of the European Parliament
MHLG	Ministry of Housing and Local Government
MINIS	Management Information System for Ministers
MP	Member of (UK) Parliament
NGO	Non-governmental Organization
NWC	National Water Council
OECD	Organisation for Economic Co-operation and Development
OJ	Official Journal (of the European Communities)
PHLS	Public Health Laboratory Service
QMV	Qualified Majority Voting
RCEP	Royal Commission on Environmental Pollution
RSPB	Royal Society for the Protection of Birds
RWA	Regional Water Authority
SAC	Special Area of Conservation
SEA	Single European Act
SEIA	Strategic Environmental Impact Assessment
SoSE	Secretary of State for the Environment
SPA	Special Protection Area
SSSI	Site of Special Scientific Interest
TEU	Treaty on European Union ('the Maastricht Treaty')
UKREP	UK Permanent Representation
UN	United Nations
UNICE	Union of Industrial and Employers' Confederations
UWWT	Urban Wastewater Treatment
WHO	World Health Organisation
WWF	World Wildlife Fund

Preface

It is a commonplace that in the course of the last 30 years British environmental policy has been deeply and irreversibly 'Europeanized' by the policies adopted by the European Union (EU). By 'Europeanized' I mean re-oriented to fit the social, political and economic dynamics of the EU. Because of Europeanization, decisions that were traditionally taken at the national or sub-national level in Britain by British politicians, civil servants and local implementing officials now involve supranational officials, European pressure groups and European politicians in a significantly more multi-levelled system of environmental decision-making.

Nowadays, there is broad agreement that Europeanization has benefited Britain by raising domestic environmental standards, addressing important cross-border problems, and making the domestic standard-setting process more formal and transparent. Yet the precise reasons for this shift – some might justifiably say 'revolution' – in British environmental governance have not been fully described, let alone explained. As a somewhat detached Member State that entered the EU with a highly sophisticated and well-established environmental management system dating back to the mid-nineteenth century, one might have expected Britain to resist or at least modulate Europeanization, to 'fit' (and hence leave undisturbed) pre-existing national practices. Britain did indeed resist the EU's involvement in what were then (and, in some quarters, still are) perceived to be overwhelmingly 'national' affairs. Nevertheless, British policy has been profoundly 'Europeanized' in the sense that EU environmental policies have altered deeply settled traditions, styles and habits of British policy-making. Today most, if not all, British environmental policy is developed by, or in close association with, the EU.

This book seeks to explore this basic paradox in modern British history, namely how and why a reluctant but environmentally advanced Member State came to be so deeply Europeanized by the 500 or so items of EU environmental legislation. It departs from existing accounts (which tend to focus on the 'top-down' impact of European political integration on national practices, rather than the constant interaction between the national and the European) by exploring the mechanisms through which national autonomy is being steadily constrained, as more and more national policy is co-determined with European bodies.

Moreover, it tries to understand the increasing inter-penetration of national and EU-level policy processes by adopting a uniquely *departmental* (or ministerial) perspective. National departments are important because they represent Britain in the EU. They negotiate EU policy with other national departments and the institutions of the EU. Between 1970 and 2000, the national Department of the Environment (DoE but, since 2001, the Department for Environment, Food and Rural Affairs, or DEFRA) was *the* primary channel of communication between EU and British policy spheres. It was the 'lead' department when most of the 500 or so EU environmental policies were negotiated, and it supervises their implementation in Britain.

Using different theories of European political integration and its mirror image, Europeanization, the chapters of this book explore the DoE's handling of the big 'history-making' alterations to the founding Treaties of the EU, namely the Single European Act (SEA), the Maastricht and Amsterdam Treaties, as well as the more mundane processes of policy evolution in the air, water, nature conservation and land-use planning sectors, in order to understand why (parts of) the British government agreed to pool so much national sovereignty in the EU, and to appreciate the long term ramifications of those decisions in Britain as well as in the EU.

We already know a lot about how the EU has affected the legal content of British environmental policy; the chapters of this book show that the EU has also hugely affected the way British policy is made. They also reveal that the increasingly deep interaction between Britain and the EU has profoundly affected the Department as well. EU policy has obviously reduced the DoE's autonomy by introducing strong, European constraints on its ability to make and implement national environmental policy. On occasions it has also generated enormous political controversy by challenging the department's policy priorities and departmental interests. Curiously, however, Europeanization has also empowered the DoE with respect to cognate national departments by increasing its bargaining power within Whitehall and giving added strength to domestic environmental protection requirements that had previously been unsystematic and weakly implemented. However, only departments that 'think (and act) European' are capable of grasping the wider opportunities presented by EU membership. When, after a long period of hesitation, the DoE finally rose to this challenge, it realized that some of the constraints imposed by membership could be more than offset by the ability to use the EU to influence environmental standards in other parts of the EU as well as internationally.

The most important and startling finding of this book is that these and many other outcomes of Europeanization in Britain were neither anticipated nor initially desired by the DoE, whose responsibility it was to manage the interface between EU and national policy. Parochial in its outlook and more economic than environmental in its values, the DoE failed to engage positively in the early, critical phases of EU environmental policy development. Rather than shaping the EU in its own image, the DoE tried to hold back the Europeanization of policy during the 1970s and 1980s by blocking EU policies in the Council of Ministers and subverting environmentally beneficial Directives at the implementation stage. The DoE also sanctioned many important EU policies in the mistaken belief that they would not unduly disrupt British environmental policy. But when these policies were eventually implemented in Britain, they were transmogrified in ways that surprised many Whitehall departments (including the DoE), but delighted national environmental groups and the European Commission. The result was Europeanization on the EU's terms, rather than Whitehall's.

In time, the political crises triggered by the growing 'misfit' between EU requirements and Britain's capacity to absorb them began to permeate through into the heart of the British administrative system, altering the DoE by making it more European and more environmental. The ensuing shifts in the DoE's structure, its departmental 'culture' and its long-term political interests, are documented in the context of the changing politics of Whitehall since 1970 and longer-term shifts in Britain's relationship with Europe and environmental policy.

This book, which draws on ESRC (R000237870) financed research, works at three different levels. It can be read as a theoretically informed analysis of the development of EU environmental policy. European scholars will find that it contains the first detailed empirical account of how environmental matters were incorporated into the three big Treaty changes, as well as the politics surrounding the negotiation of several important Directives. It can also be read by students of British politics as a detailed empirical account of the Europeanization of an important sector of British policy-making. The case studies represent the first sustained attempt to understand how the EU has profoundly affected the working methods and worldview of a large government department in Britain. However, the main argument of this book is that much more can be learnt about the contemporary architecture and functioning of European governance by studying these three interacting elements – namely European integration, the Europeanization of national policy and the Europeanization of national government – together, rather than in isolation. Studying Europeanization sheds new light on old debates about European integration in the EU. By the same token, if Europeanization research treats European integration as a given, it will fundamentally misunderstand the intimate co-evolution of national and EU policy in Europe.

ANDREW JORDAN

1
Learning to 'Think European'

[E]ach . . . department will be responsible for its own European
thinking, for knowing the European rules and for having a feel
for its aspirations; for knowing the direction in which we
should wish to see the policies of the Communities evolve . . .
This arrangement will represent a major change for Ministers
and civil servants . . . [W]hole departments must now be learning to 'think European' and to take account at all times of the
obligations imposed by Community membership.

(Sir Christopher Soames, 1972)[1]

In the last 30 years or so, a quiet but hugely important revolution has
occurred in Britain as membership of the EU (European Union) has
steadily, but profoundly, 'Europeanized' many, if not all, aspects of
domestic political life. Initially, the 'EU effect' was fairly limited even in
policy sectors such as environmental protection where the EU is now
highly active. Most people (wrongly) assumed that the EU would not
disrupt well-established British practices and academics continued to
study Britain and Europe as though they were entirely unconnected
political systems. In 1981, the Director of the Institute of European
Environmental Policy (IEEP), Nigel Haigh (1984, p.i), began to produce
the first comprehensive assessment of the EU's unfolding impact on
British environmental policy. Back then, he 'shared a view widely held –
and certainly held within the Department of the Environment – that
Community environmental policy had had little or no effect on Britain'.
As European integration accelerated in the 1980s, the EU began to affect
domestic politics in complex and often very profound ways. Some
Whitehall departments tried to attenuate the 'EU effect' by subverting
environmental Directives in a bid to hold back Europeanization. As

1

more and more scholars began to document the process and the outcomes of Europeanization, the widely held assumption that 'Europe did not matter' was first challenged and then overturned. By the mid-1980s it was clear to both scholars and practitioners alike that British policy and politics *had* been (and would continue to be) deeply affected by the EU.

Nowadays, virtually all British environmental legislation is driven by, or developed in close association with, the EU or international bodies. In 1984, Haigh (1984, pp.294, 306) shrewdly observed that: '[I]t is beyond dispute that membership of the [EU] has come to change the way an important part of British environmental policy is now thought about, is enunciated and ultimately is even put in to practice . . . [N]ational and Community policies, though distinct, cannot now be considered separately.' The British Government did not formally acknowledge the extent of this political inter-penetration until a decade later, in 1994: 'Environmental policy in the UK is now *inextricably bound up with* [EU] policy . . . Much of the UK's environmental protection legislation is now developed in common with other . . . member states' (HM Government, 1994, p.190; emphasis added).

Today, there is an extensive literature on the 'Europeanization' of British environmental policy. It argues that the EU has, *inter alia*: made UK policy more explicit; reduced the discretion traditionally enjoyed by local officials; introduced higher standards than would otherwise have been the case; created a more explicit and transparent policy framework; and introduced new policy tools and principles (Osborn, 1992; Haigh and Lanigan, 1995; Lowe and Ward, 1998; Knill, 2001). In Peter Hall's (1993) terms, European pressures have altered (though not completely overturned) the fundamental paradigms underpinning British environmental policy, the instruments used to attain policy goals and the precise setting of those instruments (Weale, 1997; Jordan, 1998b).

These impacts are well known and widely agreed upon. However, they remain just that: impacts or outcomes of European integration. The purpose of this book is to investigate how they came about: that is, to understand what actually *drives* Europeanization at the domestic level, namely the negotiation, adoption and implementation of European environmental policies (i.e., European political integration), and to analyse its impact in Britain. Until now these two processes, namely European integration (i.e., the development of common rules and institutions at the EU level) and Europeanization[2] (i.e., the process through which Member States are adapting themselves to handle the EU's growing decision-making power) have been studied separately, rather than together. There is, however, a rapidly emerging school of thought which suggests that Europeanization greatly affects subsequent phases of European policy-making by altering the very domestic terrain upon which state actors develop and articulate their preferences in relation to European integration. Existing scholarship has also represented the EU as a top-down force affecting British practices, when in fact national actors are more than capable of influencing the course of political integration in the EU, as well as the incorporation of EU rules in Britain. In other words, domestic actors do not simply 'take' Europe as they find it: Europeanization 'is a *shaped* process, not a passively encountered process' (Wallace, 2000a, p.370; emphasis added).

This book shows that Europeanization is at once an *outcome* of Britain's evolving response to European integration and an important part of the broader *process* of bringing that about. Or, as Helen Wallace (2000b, p.4) has remarked, Europeanization and 'domestication' are a 'two way street in which the dynamics depend on the interplay between the two'. In order to capture the 'rebound' effect we need to understand how Europeanization is 'domesticated': that is, framed and influenced by domestic institutional and political factors (Wallace, 2000a, pp.369–70).

Europeanization and domestication

A departmental perspective

This book explores the co-evolution of European and British environmental policy in the period 1970–2000. It departs significantly from existing accounts by focusing on the activities of the national Department of the Environment (DoE). Why should scholars pay more attention to particular departments of state if most of the more obvious impacts of Europeanization are on national policy? Departments are important because they are *the* primary channels of communication between national and European political spheres. As the primary units of national government, they develop (and are responsible for supervising the implementation of) the bulk of public policy, including, crucially, EU policy. As the chief *mediators* of the reciprocally interlinked processes of European integration at the European level and Europeanization at the domestic level, they are in an ideal position to shape (and be shaped by) both in pursuit of their departmental interests.

Unfortunately, the existing academic literature has barely begun to uncover what might be termed the inter-departmental politics that drive and are driven by Europeanization and European integration (but see Pehle, 1998). In fact, it was only relatively recently that the DoE itself realized the importance of Europe and took purposeful steps to improve its understanding and handling of European policy. The main thrust of this book is that departments of state do *matter* in shaping the continuing inter-penetration of domestic and European politics. However, their influence is greatly constrained by contextual factors, of which some are domestic (e.g., public attitudes to the environment and Europe, pressure from cognate Whitehall departments, etc.) while others are European (e.g., pressure from supranational actors such as the European Commission and other Member States). So, although the DoE forms the main analytical focus of the book, individual chapters carefully relate its behaviour to the broader structures of national and EU policy-making. These include cognate 'environmental policy' departments in Whitehall, notably industry, agriculture and, to a lesser extent, transport, the 'core executive' (i.e., the Cabinet, the Foreign and Commonwealth Office, or FCO, and the Prime Minister's offices),[3] and long-term shifts in Britain's economic, social and political relations with continental Europe.

Processes and dynamics

Throughout the history of the EU, states have tried to shape European rules to ensure they are aligned with their own national approaches and practices. By working to ensure a 'goodness of fit' (Green Cowles, Caporaso and Risse, 2000) between the two, states hope to reduce adjustment costs, achieve 'first mover advantages' and thereby reduce political and legal uncertainty. The 'regulatory competition' (Héritier *et al.*, 1996) between the 15 Member States to set the 'rules of the game' at the European level defines the scope of European integration. Crucially, this process inevitably creates instances of institutional 'misfit' when European requirements conflict with the way in which states have traditionally organized their domestic environmental affairs (i.e., the structures, style and philosophy of national policy). It is commonly argued that these 'misfits' give rise to Europeanization (Green Cowles, Caporaso and Risse, 2000). The logical implication of this argument is that more proactive states can forestall Europeanization by exporting, uploading or projecting (Bulmer and Burch, 1998) their preferred national policy approach to the EU. After all, if (as has often been the case with the more environmentally progressive states such as The Netherlands – Liefferink, 1996 – and Germany – Weale, 1992) European rules are based on the core features of national policy, the misfit is likely to be low and the degree of Europeanization correspondingly weak. By contrast, states that consistently download EU policies which are modelled on alien institutional systems will find themselves under European and domestic pressure to fall into line. The gradual accumulation of misfits will eventually produce serious implementation problems, significant political crises and, possibly, sudden domestic transformations (Green Cowles, Caporaso and Risse, 2000, p.8).

There are two problems with this analytical approach. First, it implies that states are either uploaders (policy-shapers) or downloaders (policy-takers) of EU policy. In practice, even the most proactive national departments struggle to satisfy their interests consistently because EU policy is really a complex amalgam of different national approaches (i.e., a 'patchwork' (Rehbinder and Stewart, 1985, p.254)). Consequently, no single state can hope to be perfectly and consistently aligned with EU requirements. Even though the Commission often bases its proposals on pre-existing national practices, there will, by definition, *always* be 'misfits'. The real puzzle, therefore, is to understand how these misfits arise and are managed: that is, how individual countries (represented by their lead departments) manage the continuing, multi-level politics of integration and Europeanization which encompasses the negotiation of policy in Brussels through to its implementation in domestic settings.

Second, notions such as 'fit' and 'misfit', 'download' and 'upload' usefully convey the dynamism of European policy-making, but they lack human agency. More often than not, the 'taking' and the 'shaping', and the 'uploading' and the 'downloading', are done by national departments of state. Departments are, in effect, the main policy 'fitters' in the EU. How these organizations go about resolving the 'misfits' between the logics of EU and national policy is of obvious importance, not only to scholars of the EU but also policy practitioners in Brussels and Britain. In one sense, this book offers a detailed empirical example of what Keohane and Nye (1972, p.376) famously termed the 'internationalization of domestic politics' (i.e., 'outside in') and the 'domestication of international politics' (i.e., 'inside out'), by studying integration and Europeanization together, as a single iterative and interactive process which spans different levels of governance, but is channelled through national departments.

The DoE (or DEFRA as it is now known – see below) is in an especially privileged position because many (though by no means all) of the important, day-to-day connections between the EU and Britain pass through its offices. It can: communicate (i.e., 'upload') policy ideas to the European Commission as it develops new policy proposals; attend the Commission working groups which define the technical content of proposals; input – via UKREP (the British permanent representation in Brussels) – to the inter-governmental committees which negotiate the final content of proposals; indirectly influence the views of the European Parliament's committees which pass judgement on Commission proposals; directly articulate its position in the Environment Council, the EU's chief environmental policy-making body, where final political agreement is reached; and modulate EU policies as they are implemented in Britain. Other actors will, of course, be involved in all these activities, but not nearly to the same extent or intensity as the DoE.

In order to achieve a goodness of fit (or, conversely, avoid a 'badness of fit') between national and EU rules, the DoE must work hard in and across all these policy venues. It can set political agendas or define the scope of issues before they become solidified in Commission proposals. It can work with UKREP in the various committees of the Environment Council to erode or modulate policies before they are downloaded from the EU. In respect of the Commission's 'comitology' committees that oversee the implementation of Directives (Weale *et al.*, 2000, pp.90–1), it could mean finding the right national scientists and/or technical experts to project the national interest. It may also mean developing constructive relationships with non-state actors such as the European Parliament (EP), the Commission, European and national pressure groups in order to shape the endgame of negotiations in the Council itself. Finally, if 'projection' fails and uncongenial EU policies have to be downloaded, as the lead department, the DoE is primarily responsible for finding ways to avoid politically and economically costly misfits, while at the same time ensuring that Britain complies formally and practically[4] with EU requirements. One way of achieving this is to *modulate* the overall scope and impact of Europeanization by means of slow or partial implementation.

Context and perspective

Evidently, behind the concept of 'misfit' lies a dimly lit sequence of interactions, spanning multiple levels of governance and involving many actors. Yet, apart from a handful of studies dating back to the accession period, the inter-departmental politics of integration and Europeanization have not been studied systematically (but see Jordan, 2001b, and J. Smith, 2001). Why is this? An important factor was the widely held belief that EU membership would not disturb domestic policy-making in Whitehall. For example, Wallace and Wallace (1973, p.261) predicted a process of 'gradual adaptation rather than . . . radical change' and noted a 'mood . . . of confidence in the capacity of existing structures and procedures' to cope with European demands. Academics too, assumed that European integration would put *inter*-state relations on a new footing but leave *intra*-state politics relatively untouched. The enduring popularity of highly aggregated models of the state (which assume that national politicians and bureaucrats work together to achieve a set of exogenously derived 'state preferences') is one manifestation of these attitudes (see below). Another is the dominance of the 'Whitehall Model' of the British state (M. Smith, 1999), which has served to perpetuate the myth that national departments are sovereign and largely unaffected by international influences.

The purpose of this book is to explore the role that inter-departmental politics plays in shaping the multi-level interaction of European integration and Europeanization. In particular, it investigates *when, why* and *how* the DoE 'mattered' in the continuing co-evolution of EU and British environmental policy. Standing as it does at the interface between the British and EU political systems, one would expect it to have had some impact on the interplay between European integration and Europeanization within this particular policy sector. Existing scholarship approaches its role from two very contrasting theoretical positions. The first is state-centric in the sense that it assumes integration and Europeanization are fairly predictable processes, directed by states to achieve national political and economic objectives arrived at through domestic processes of preference formation. Central governments are regarded as monolithic but internally highly coordinated entities, dealing with discrete packages of policy: environment, agriculture, the economy and so on. On this view, individual departments do not 'matter' that much, being little more than passive weathervanes, which are blown one way or the other by domestic political and economic forces (Moravcsik, 1998).

The second presents European policy-making as a much more unpredictable, open-ended and contingent process. These characteristics are not necessarily uncovered by state-centred studies, which typically select single, set-piece events or items of legislation and study them over fairly short periods of time. More 'process-based' theories argue that when Europeanization and integration are studied *together* as a gradually unfolding *process* of change, the constraints on state capacity become more apparent. They suggest that 'goodness of fit' is a constant and dynamic *process* of mutual adaptation, which is affected by (and, in turn, affects) the goals pursued by different actors through a series of feedback loops. Departments (rather than 'the state') 'matter' because most of the day-to-day policy decisions relating to integration and Europeanization are bargained over in highly specialized policy networks in which they and few (or no) other British departments participate (Peterson and Bomberg, 1999). Therefore, departments 'matter', but not necessarily in the highly predictable and instrumental manner posited by state-centric theories. In fact, as national policies are gradually Europeanized by the EU, advocates of process-based theories would expect national departments to adjust their own internal identities and preferences to reflect the growing ascendance of European institutional 'logics' of appropriate behaviour.

Outline of the book

The remainder of the book proceeds as follows. Chapter 2 describes the history of EU environmental policy, outlines the machinery of government in Whitehall which coordinates the national input to European policy, and describes the DoE's response to the Europeanization of environmental policy-making since 1970. Chapter 3 provides a more detailed exposition of the two theoretical positions and pinpoints the assumptions they make about the role of national departments. Chapters 4–10 examine different empirical cases of European environmental policy-making, the first three 'major' and the remainder more 'mundane', through the lenses of the two theories. Each chapter analyses how the aims of British policy were forged in the conflict between the DoE, cognate departments and the core executive. These aims relate both to the preferred extent of European integration as well as its eventual effect (Europeanization) on Britain. The chapters conclude by drawing on the two theoretical positions to assess the extent to which Britain's aims were realized or not. Recall that state-centric theories expect outcomes to be fairly predictable and consistent with 'state' preferences, whereas process-based theories paint a much more open and unpredictable picture. The final chapter draws together the threads of the argument and assesses the value of a departmental perspective. The remainder of this chapter unpacks the term Europeanization and identifies three aspects of departmental behaviour that deserve particular attention. Then it justifies the exclusive focus on Britain and explains the rationale for the case selection.

Coping with Europeanization

The phrase 'think European' came into popular usage around the time of Britain's entry into the then European Economic Community (EEC) in 1973. Having surveyed the way other Member States managed their European affairs, Prime Minister Heath decided not to create a European Ministry but instead to devolve European work to line departments, governed by central coordination mechanisms overseen by the Cabinet Office and the FCO. He hoped that this would encourage officials to treat EU policy work as though it were domestic policy work, and not some disconnected aspect of international affairs. It is worth noting that the advocates of EEC membership always saw 'Thinking European' as being more than just a process of administrative adjustment within Whitehall. Heath, for example, regarded Europeanization as a trans- forming and modernizing force, which would 'uncongeal the attitudes, habits and expectations of the British' (*Guardian*, 2 June 1971).

Today, the need always to 'Think European' is much more widely rec- ognized in Whitehall than it was even as recently as the mid-1980s. In the early 1990s, the then head of the civil service, Sir Robin Butler, remarked that Europeanization: 'will certainly demand a more outward looking civil service, with greater ability to work with our European partners, to deal in their languages, to have a feel for their cultures, to understand their aspirations and their administrative methods' (quoted in Drewry, 1995, p.473).

More recently, the Government issued a *Guide to Better European Regulation*, which recognizes the intimate link between integration and Europeanization. It recommended that civil servants improve their under- standing of the EU policy process, coordinate with their colleagues in cog- nate departments, cultivate links with their opposite numbers in other states and European institutions, and sharpen their negotiating skills, in order 'to ensure that EC [European Community] legislation is of the highest quality [and] bring[s] benefits to all of Europe' (Cabinet Office, 1999, p.1).

In spite of its popularity, the precise meaning of the term 'European- ization' remains contested. Some have chosen to define it as 'the emer- gence and development at the European level of distinct structures of governance' (Green Cowles, Caporaso and Risse, 2000, p.2), but this risks eliding Europeanization with the source of change: European integration. Then there are analysts such as Börzel (1999, p.574) who define it as the *impact* of European policies at the national level. On this view, Euro- peanization concerns the process through which European integration penetrates and, in certain circumstances brings about adjustments to, domestic institutions, decision-making procedures and public policies.

Like Börzel, this book also interprets Europeanization as a cause of impacts and adjustments at the national level, while at the same time uncovering the origins of those impacts in the processes of political inte- gration at the European level. The focus of the chapters is on investigating the impact of the EU in a number of sub-areas of British environmental policy, as well as on the DoE itself. Throughout the 1970s and 1980s, Britain fought doggedly to preserve its pre-existing approach to regulating environmental problems in the teeth of pressure from the EU to adopt more continental approaches (Golub, 1996). In so doing, Britain gained a not entirely undeserved reputation for being 'The Dirty Man of Europe' (Rose, 1990).[5] We now know that Britain's strategy of trying to thwart Europeanization by resisting integration was relatively unsuccessful. Eventually, EU policies had to be downloaded, resulting in a deep-seated and wide-ranging Europeanization of national policy. This book investigates this puzzling turn of events by examining Britain's behaviour in a broad cross-section of environmental policies throughout the period 1970–2000. Crucially, it seeks to understand Britain's negotiating positions in these areas by looking at the political relationship between different domestic departments and the EU. In so doing, it investigates the origins of the DoE's policy preferences. Were they, for instance, a reflection of wider British societal views on Europe and the environment, or the Department's own, departmental interests? Was the DoE intrinsically 'awkward' in the way it negotiated with other national environment departments, or was it simply arm-locked by other Whitehall departments into carrying their sceptical preferences to Brussels? Was its 'awkwardness' a reflection of the style in which it operated in Europe, or the position it adopted on particular issues having consulted other departments (Weale *et al.*, 2000, p.97)? Finally, how did the DoE interface with other departments when 'misfits' were diagnosed? Indeed, how good was it at anticipating and resolving misfits?

In order to appreciate the co-evolution of integration and Europeanization, we must analyse the long-term impact of the EU policies that were finally adopted to assess the overall extent of Europeanization. Golub (1996; 1997) adopts the language of state-centric theory when he implies that the British *Government* was well organized, knew and understood what it negotiated during the integration process, and, with a few exceptions (e.g., some aspects of water policy), remained fully in control of Europeanization. He concludes:

> a plethora of [EU] laws imposed no tighter standards or new regulatory costs on domestic groups. In most cases national legislation was already in place or under consideration, or [EU] regulatory standards were already met without explicit domestic legislation. The combination of vague provisions, derogations and loopholes makes it difficult to characterise [EU] environmental law as the source of a steady ratcheting effect in the member states . . . All the evidence indicates that powerful industry/producer groups and government representatives defended predictable national positions [in the EU].

(Golub, 1997, p.18)

However, he does not really open up the 'black box' of the state and investigate the DoE's handling of European affairs or explore its relationship with other national departments, and neither does he adopt a sufficiently long time span to be sure that the implementation of the EU policies actually generated the impacts (i.e., the Europeanization) that British negotiators in the DoE expected. If they did, then perhaps the state-centric view is worthy of further examination. But if integration generated unforeseen and unfavourable outcomes, then perhaps state control is more circumscribed than state-centric theorists claim, and other theoretical models are more appropriate.

Analytical themes

The remainder of this book examines three distinct but highly interrelated aspects of the DoE's handling of the relationship between European integration and Europeanization. The first concerns its overall contribution to European integration: that is, to what extent has it facilitated and/or retarded the development of rules and rule-making structures at the European level? The second addresses its contribution to the Europeanization of national environmental policy: to what extent has the DoE actively shaped Europeanization? The third concerns the DoE's own, internal response to the Europeanization of policy, both in terms of its organizational structure and philosophy, and the way it interfaces with other actors.

The Department of the Environment and the politics of European integration

Insofar as European scholars examine national departments, it is usually to answer one of two main questions: (1) why do states decide to pool state sovereignty; and (2) to what extent do they retain control over European rule-making? The DoE could conceivably decide to pool national sovereignty as a means of levering a commitment to stronger environmental policy from cognate departments. State-centric theorists fiercely deny this possibility, arguing that departments typically enjoy very little leeway to negotiate outside the framework of exogenously determined national preferences. Process-based theories, on the other hand, claim that individual departments do have their own independent role interests, shaped by the endogenous dynamic of European integration. That is to say, integration triggers processes of Europeanization at the national level, which then affect the way departments feed into the next round of European negotiations. Sbragia (1996, p.247) conjectures that 'the dynamic [of EU environmental policy] is one in which countries with relatively little environmental interest are drawn into a policy-making process in which ministers of the environment play a more prominent role than they may in the national government'.

How likely are we to find the DoE playing multi-level politics in Europe to advance its domestic departmental interests? The DoE was in the second wave of government departments affected by Europe, some way behind the Ministry of Agriculture, Fisheries and Food (MAFF), the FCO and the Department of Trade and Industry (DTI), but ahead of much more domestically focused departments such as health and home affairs. However, unlike the vanguard departments, the DoE joined when the EU was still in the very early stages of developing an environmental policy (see Chapter 2). In theory, this handed the DoE a golden opportunity to shape the rules of the EU game as they were being developed. As the first integrated environmental department in the world and the guardian of one of the oldest systems of environmental management in the industrialized world, the DoE was in a good position to 'shape' rather than 'take' EU environmental rules. Unlike the most domestically focused departments, the DoE should have been well disposed to Europe, because environmental problems often transcend borders and are, in principle, highly amenable to 'European' solutions.

Yet existing accounts characterize Britain (and by implication the DoE) as a taker (i.e., a 'downloader') of other countries' policy proposals rather than a net exporter (i.e., an 'uploader') of ideas (Lowe and Ward, 1998, p.287). Current scholarship does not, however, provide a clear and convincing explanation for Britain's behaviour. Was it, for instance, because of pressure exerted by older and politically more powerful departments such as trade, agriculture or foreign affairs? Or was the DoE inherently suspicious of the EU? This book takes a detailed look at the bureaucratic structure of the DoE, its standard operating procedures and departmental 'culture' – that is, its own developed sense of mission – and the 'appreciative systems' of its staff, to see if they yield clues.

The Department of the Environment and the Europeanization of national policy

It is a commonplace that British environmental policy has been deeply Europeanized, but it is desperately unclear whether this outcome occurred because or in spite of the DoE. This puzzle cuts across and is directly relevant to the literature on Europeanization and European integration. On the face of it, Golub's (1996; 1997) claim that states control European integration sits rather oddly alongside the deep and lasting

Europeanization of policy detected by most other scholars. One possible explanation is that the DoE wrongly assumed it was in control of the Europeanization of policy when it downloaded European rules (i.e., that it miscalculated the long-term domestic effect of European integration). Another is that the DoE deliberately signed up to open-ended environmental commitments as a means of surreptitiously strengthening domestic environmental protection. By concentrating on the adoption of European rules (i.e., integration) and neglecting their impact at the national level (Europeanization), Golub is unable to adjudicate between these possible explanations. By contrast, the chapters of this book look at the *co-evolution* of integration and Europeanization over a 30-year time period. This involves carefully tracing a series of decisions through the full policy cycle from agenda setting to implementation, in order to examine the extent to which they followed predictable or unpredictable pathways. Of course our two theoretical approaches lead us to expect very different outcomes: state-centric theories maintain that departments are capable of anticipating and in turn managing integration and Europeanization; process-centred approaches disagree, given the complexity and unpredictability of integration and Europeanization.

The Europeanization of the Department of the Environment

Finally, relatively little is known about how individual Whitehall departments are responding to European pressures. There is flourishing interest in the Europeanization of inter-departmental coordinating mechanisms (Kassim, Peters and Wright (eds) *et al.*, 2000), but it tends not to reach into the internal aspects of departmental life to investigate how far the EU is affecting interests and identities (but see Buller and Smith, 1998). Currently, there are two broad schools of thought, though neither is built on a substantial empirical base. The first adds weight to the state-centric claim that states are generally unaffected by European integration, although many of the researchers involved would resist the label 'intergovernmental'. It suggests that the Europeanization of departmental structures in Britain has been relatively limited. On the whole European policy pressures have been absorbed relatively smoothly into the 'administrative logic' of Whitehall departments with very little obvious disruption (Bulmer and Burch, 1998, p.606). When compared to the 'deep' Europeanization of national policy and politics, the 'EU effect' on Whitehall has been much less dramatic and mostly incremental:

Membership has brought new issues onto the agenda, altered the terms of the debate concerning established issues, given whole areas of policy a European dimension, required the development of new expertise on the part of officials and ministers, involved extensive and intensive negotiations with EU partners and raised significant problems about policy presentation and party management. Yet *at the level of machinery, governmental structures and procedure, the impact of Europe has been far less evident* . . . The significant challenges of European membership have been characterised predominantly by a process of slow and steady adaptation.

(Bulmer and Burch, 1998, p.624, emphasis added)

Moravcsik (2000, pp.286–8) believes the real question is not whether state structures are transformed or not, but how far Europeanization achieves national interests. In a sense his interpretation of Europeanization is the mirror image of the state-centric view of European integration (i.e., predictable and essentially state-dominated).

Another school of thought holds that the Europeanization of departments has been much more substantial and long lasting. It claims that Europeanization not only shapes the administrative procedures and conventions of departmental life, but also the very preferences and identities of individual parts of the state, including departments. In keeping with more process-based theories, it conjectures that complex policy feedback effects engendered by integration fundamentally transform the actors involved in European policy processes (cf. Heath's comments above). At its most strident, the 'transformation of the state' argument suggests that: 'as a process, European integration has a *transformative* impact on the European state system and its constituent units . . . it is reasonable to assume that in the process agents' identity and subsequently their interests have equally changed' (Christiansen, Jorgensen and Weiner, 1999, p.529; emphasis in original). The chapters of this book investigate the validity of these two assertions in relation to the environmental relations between Britain and the EU since 1970. They analyse the Europeanization of the DoE by looking at long-term alterations to its internal institutions (procedures and rules), preferences and identity.

The case studies

Where should analysts look to find the empirical evidence to test our two theories? There are two main types of decision in the EU: 'big' or 'history-making' alterations to the founding Treaties of the EU, and 'smaller', 'day-to-day' decisions that make up the 'everyday grind' of initiating and implementing policies in particular sub-areas of activity. There were four major changes to the founding Treaties in the period 1970–2000: the 1987 Single European Act (Chapter 4), the 1993 Maastricht Treaty (Chapter 5), the 1999 Amsterdam Treaty (Chapter 6) and the Nice Treaty. At the time of writing, it is still too early to assess the full impact of the most recent changes (which were adopted in Nice but await ratification) so they are excluded from the analysis (but see Jordan and Fairbrass, 2002).

Numerous studies have demonstrated the tendency for integration to proceed informally in the period between these major Treaty alterations. State-centric theorists dismiss these claims as 'anecdotal' because they cover policies (e.g., regional policy) that were designed to shift power downwards to sub-national government (Moravcsik, 1994, p.53). This book therefore takes up Moravcsik's challenge to sample 'essential structures [i.e., British central government] and major policy decisions [i.e., the Treaty changes]' (1994, p.52) through the full policy cycle. Thus each case study chapter considers the origins, negotiation and longer-term consequences of the intergovernmental conferences (IGCs) that culminated in the Treaty changes in relation to EU environmental policy. In other words, they trace the changes from the construction of state (or, in process-based terms, 'departmental') preferences through to outcomes, including any unintended consequences. As will become apparent in Chapter 3, a 'process-tracing' approach can generate insightful findings: because things operate in a particular way today, this does not necessarily imply that they were designed to do so by their initiators; they may be the product of unforeseen or unintended consequences.

Turning to the small decisions, Chapters 7–10 examine how the networks of policy expertise centred on different parts of the DoE processed secondary legislation in four sub-areas of environmental policy, namely: water pollution (the Bathing Water – EC/76/160 – and Drinking Water – EC/80/779 – Directives); biodiversity protection (the Directives on wild birds – EC/79/409 – and natural habitats – EC/92/43); air quality (smoke and sulphur dioxide – EC/80/779 – and integrated pollution prevention and control, or IPPC – EC/96/61); and land-use planning (environmental impact assessment, or EIA – EC/85/371 – and strategic environmental impact assessment, or SEIA – EC/2001/42). Although selection bias is difficult to avoid when there are over 500 items of EU environmental policy, these six Directives cover a good cross-section of environmental media, periods of regulatory activity and modes of intervention. Crucially, they cover a much wider range of policy areas than the two (air and water) covered by Golub (1997) in his assessments. The case study approach does, of course, have its limitations (Keohane, King and Verba, 1994). It is particularly important to ensure that the chosen cases are properly situated, by which I mean placed in their appropriate historical and organizational context. To that end, Chapter 2 describes the wider, historical evolution of EU and British environmental policy since 1970. Each case study chapter then takes this description down a level to the historical development of policy within each specific policy area, carefully comparing national policy before and after the EU's involvement to gauge the overall extent of change. According to Börzel and Risse (2000), the EU can lead to small (absorption), medium (accommodation) and large (transformation) policy changes at the national level (see Table 1.1). However, the purpose of this book is not only to assess the degree of domestic policy change (Europeanization, measured in terms of policy paradigms, policy tools and the calibration of those tools; see above and P. Hall, 1993), but also to contribute to the debate about European integration by examining the extent to which that change was actually intended by (different parts of) the state.

The case studies draw upon a wide variety of documentary sources, including government and Parliamentary reports, the Explanatory Memoranda that British officials append to Commission proposals, DoE internal staff notices and press releases dating back to 1970, DoE MINIS[6] returns and internal files to which the author was granted privileged access. However, some argue persuasively that Europeanization operates in a much more subtle way 'in the bowels of British administration' and can-

Table 1.1   Degrees of domestic policy change

	Extent of policy 'misfit'	*Amount of domestic change*
Absorption	Small: EU and national policy similar	Small: States are able to incorporate/domesticate EU requirements without substantially modifying national policies
Accommodation	Medium: EU and national policy differ	Medium: states accommodate/mediate EU requirements by adapting existing policy while leaving its core features intact
Transformation	High: EU and national policy markedly different	High: domestication fails; states forced to replace or substantially alter existing policy to satisfy EU requirements

Source: based on Börzel and Risse (2000).

not fully be explored using documentary sources (Christoph, 1993, p.534). In order to penetrate the 'work-a-day world' of civil servants – that is, the culture, rules and values which bind them to other actors in policy networks and influence the way they process and interpret information – 'we must begin by seeing the world through their eyes' (Heclo and Wildavsky, 1981, p.lxvii). Although time consuming, elite interviews help to reveal the subjective underpinnings of policy-making. This book draws upon over 60 interviews with politicians, civil servants, environmental activists and EU officials, which were conducted between 1998 and 2001.

Analytical parameters

Before moving on to consider the historical co-evolution of EU and British environmental policy, it is important to clarify the analytical limits of the book. First, the account centres on events in Britain, and specifically England and Wales. Environmental policy in Scotland and Northern Ireland is organized rather differently, and is not addressed directly by this book.

Second, the chapters concentrate on the DoE in the context of its relationship with the rest of Whitehall. Why does the DoE's role merit closer attention when there are currently 15 'environmental' departments in the EU? Although this book is not strictly comparative, there are good reasons for treating the DoE and Britain as a 'critical case' of Europeanization. Britain has traditionally approached European negotiations with great scepticism. With a strong, centralized bureaucracy and tough internal co-ordinating mechanisms, Britain also has a reputation for administrative excellence in Europe, which is the envy of many other states (see Chapter 2). As one of the three largest states in the EU, we would not therefore expect DoE officials unwittingly to lock themselves into an open-ended process of political integration, agree to policies whose effects (Europeanization) might be uncertain or unpredictable, or use the EU to gain leverage over other departments by acting unilaterally in Europe. Therefore, to find evidence of these would strike a blow to state-centric accounts of the EU.

Third, this book examines the Europeanization of British *environmental* policy from a departmental perspective and consequently it concentrates mainly, though by no means exclusively, on the 'environmental' parts of the DoE, broadly defined to include pollution control (the responsibility of the DoE's Environmental Protection Group, or EPG), land-use planning and habitat conservation, rather than the sections dealing with housing, local finance and (at various times) transport. The DoE is not a monolithic entity. Individual chapters reveal that European policy has intruded into the bowels of the EPG, whereas the EU's impact on some of the 'non' environmental parts of the department is somewhat lighter and more recent. On a continuum of high to low EU involvement, water and air pollution policy and regional policy would be at one end, whereas land-use planning, public procurement, local government finance and housing would be at the other (DoE, 1993c, p.7).

It is also worth noting that both the structure and the name of the British environment department have changed substantially in the period 1970 to 2000. The DoE was created by the Heath Government in 1970, and was the world's first integrated environmental department (see Chapter 2). In 1997, the DoE was re-integrated with the former Department of Transport to form the Department of the Environment, Transport and the Regions (DETR). In 2001, Prime Minister Tony Blair split this mega-department and fused the environmental parts of the DETR (i.e., the EPG but not planning) with MAFF to form the Department for Environment, Food and Rural Affairs (DEFRA). For convenience, the term DoE is used throughout this book, although the story told relates mostly to specific parts of the EPG (e.g., air, water, etc.) as well as to land-use planning.

Fourth, wherever possible in the text, attempts are made to disentangle the 'European' drivers of change (i.e., Europeanization) from the 'non' European drivers (e.g., national or international policies, etc.), otherwise there is a danger of ascribing everything (or nothing) to the EU's involvement. Finally, a note about the various terms used to describe the EU. Until the ratification of the Single European Act (SEA) in 1987, the European Union was known as the EEC. The SEA officially re-christened it the European Community (EC), a term which remained in popular use until 1993, when the Maastricht Treaty created the EU with a 'three-pillar' structure. Strictly speaking, environmental policy is still made within the first 'pillar' of the EU (i.e., the EC). For the sake of convenience and consistency, the term 'EU' is used throughout this book, even though, strictly speaking, the earlier phases (i.e., 1970–87) concern the EEC. Just to confuse matters still further, many of the legal articles of the Treaty referred to in the text were re-numbered by the 1999 Amsterdam Treaty. So, for example, the environmental Articles 130r, s and t of the SEA are now Articles 174–6 of the Amsterdam Treaty, and so on. For the sake of simplicity and consistency, the text uses the *pre*-Amsterdam system for events occurring prior to Amsterdam.

2
European Union Environmental Policy and Britain

At its founding in 1957, the EU had no formal environmental policy and no environmental bureaucracy. The European *Economic* Community (EEC) was primarily an *intergovernmental* agreement between six states (none of whom had environmental ministries) to achieve social and economic prosperity. Today, the EU has some of the most progressive environmental policies of any state in the world, although it is not actually a state. Moreover, contemporary EU environmental policy adds up to considerably more than the sum of national environmental policies. In the EU, national environmental policies are no longer legally or politically separate from EU environmental policy: in a word, they have been deeply Europeanized as a result of their interaction with EU policy-making. This chapter describes the history of EU environmental policy, then explains how the British government has responded to the continuing Europeanization of national policy-making. The final section traces the DoE's historical relationship with EU environmental policy, drawing out the links between European integration and the Europeanization of British policy.

The evolution of European Union environmental policy

Birth

When Britain joined the EEC in 1973, the environmental *acquis* (i.e., the total stock of EU policies, rules and guiding principles) comprised a very small number of laws whose primary aim was to safeguard human health and remove internal barriers to trade. In stark contrast to today, there was very little political support for protecting the environment for its own sake. Throughout the 1970s and early 1980s, items of EU environmental policy had to be unanimously agreed by the Environment Council on the basis of proposals submitted by the Commission. It was only after the surge of environmental awareness in the late 1960s and early 1970s, which culminated in the 1972 United Nations (UN) environment conference in Stockholm, that the EEC started to address the environmental agenda seriously. A few months after Stockholm, European leaders met in Paris to initiate policies in areas such as the environment, which would give the EU 'a human face'. The Commission subsequently drew up a very short and simple Programme of Action on the Environment in 1973, which marked the start of a coordinated and purposeful *European* environmental policy.

Throughout the remainder of the 1970s, individual items of EU policy were adopted in a very slow and *ad hoc* manner, according to the whims of particular states. In the first decade, no single state or coalition of states consistently dominated the EU by uploading its national policy (Rehbinder and Stewart, 1985, p.262). In many crucial respects the EEC functioned just like any other international organization, allowing states selectively to download – or what the DoE's Chief Scientist in the 1970s, Martin Holdgate (2000), terms 'cherry pick' – policies from the *acquis* to fit their economic and political circumstances. Not surprisingly, very few people expected Britain (or, indeed, any other Member State) to be deeply Europeanized by the EU.

Having formally established an environmental role for the EEC, European political leaders' states moved on to address other issues such as the economy. In 1973–4, an economic recession coincided with declining levels of political support for integration, bringing integration to a virtual standstill in many sectors. Politically speaking, the decade from the mid-1970s to the launch of the single market in the mid-1980s (Chapter 4) was a lost or 'stagnant epoch' in the history of the EU (Weiler, 1991, p.2431). But in more 'technical' areas, such as the environment, European integration continued to inch forward, driven by policy entrepreneurs such as the Commission:

> The momentum [of political integration] was directed at a range of ancillary issues, such as environmental policy, consumer protection, energy, and research, all important of course, but a side game all the same. Yet, although these were not taken very seriously in substance (and maybe because of that) each . . . represented part of the brick-by-brick demolition of the wall circumscribing Community competences.
>
> (Weiler, 1991, p.2449)

Although each individual item of legislation was relatively unimportant, together they amounted to a substantial body of EU legislation and policy, which is now deeply Europeanizing national policy. The case studies which follow investigate how far the process of accumulation occurred at what Weiler terms a 'subterranean' level (1991, p.2408) in the bureaucracies of the EU and the Member States, well away from the spotlight of political attention. His point (above) about environmental policy not being taken 'very seriously' in national capitals also raises important questions about the extent to which national departments of state actually controlled the process of integration (and, eventually, Europeanization). Again, it suggests that we should, as explained in the previous chapter, take a closer look at the interdepartmental politics arising from Whitehall's handling of European integration and its domestic consequences.

In comparison to national environmental departments, the Commission's bureaucratic capacity to effect huge change was incredibly tiny: in the mid-1970s, just fifteen staff worked in its environmental unit (the Environment and Consumer Protection Service, or ECPS). The chapters of this book show how the Commission learnt to overcome these constraints by working in politically 'unimportant' areas such as bathing water (Chapter 7) and wild bird protection (Chapter 8). Here, it found it could operate relatively unsupervised by pressure groups, national politicians and government officials. With some notable exceptions, democratic oversight by the European Parliament and national parliaments was weak. National pressure groups concentrated on domestic affairs because that was where the locus of environmental policy resided. In fact, the very first pan-European environmental group, the European Environmental Bureau (EEB), was not formally established until 1974. Other actors took rather less notice of the EU than they do today and '[e]arly . . . Directives . . . owed their existence to their lack of conflict with the interests of the established and powerful [Commission] Directorates' (Haigh and Lanigan, 1995, p.22). Strictly speaking, the EU was acting beyond its legal competence by adopting environmental measures, but the 'deficiencies of the legal basis . . . were compensated by the political will of the member states' (Rehbinder and Stewart, 1985, p.246). Every now and again a particular proposal would provoke conflict between states (e.g., EIA: see Chapter 10), but otherwise the EEC remained a very weakly developed polity. The pursuit of political 'integration by stealth' (Weale *et al.*, 2000) was not simply the Commission's coping strategy; the founding fathers of the EU firmly believed that integration would occur more quickly if the Commission focused its energies on building transnational cooperation in areas of 'low politics' such as the environment, in the hope that political support would then 'spill over' into more politically sensitive areas such as the economy and national defence. In the early years of EU environmental policy, integration 'by stealth' (the Monnet method) succeeded brilliantly because, by the early 1980s, the EU had adopted around 60 environmental laws (see Table 2.1).

Adolescence

In the 1980s, this technocratic form of governance gradually gave way to a more diffuse and pluralistic web of activities centred on a set of trilateral links between the Council, the Commission, and the EP's Environment Committee. The two major driving forces for this change were the upsurge in political concern for the environment in countries such as The Netherlands and Germany, and a growing support within most European national capitals (including London) for a shared, *European* response to mutual concerns such as economic competitiveness in world markets. In the early 1980s, what had been a trickle of environmental legislation turned into a stream as the EU underwent a relatively rapid and profound transformation, which culminated in the signing of the Single European Act (Chapter 4). Very quickly, matters that had previously been contained in discrete intergovernmental committees began to enter the political mainstream, energizing national and international pressure groups, Europeanizing national practices, and exciting public interest. In 1981, the Commission increased its bureaucratic capacity to draft still more proposals by creating a separate environmental Directorate-General. By 1987, the number of EU environmental measures had climbed to over 200. Moreover, many of the new policies – which related, *inter alia*, to seals, natural habitats, sewage treatment, genetically modified organisms and climate change – went

Table 2.1   The expansion of EU environmental policy in different time periods

	1958–72	1973–6	1987–92	1993–5	1995
No. of laws adopted*	5	118	82	60	5
Average no. of laws adopted p.a.	0.3	8.4	13.7	20	5
Average no. of new and amended laws adopted p.a.[†]	0.6	13.9	32	48	28

*Regulations, Directives and Decisions only.
[†]Including amendments and elaborations.
Source: Based on Zito (2002), p.160.

well beyond what would be strictly necessitated by a concern to ensure a single market.

What factors provoked this sudden step change? The first was concerted political pressure from greener Member States such as Germany, The Netherlands and Denmark, who sought to upload their high environmental standards to the EU. At this time, Britain was often to be found blocking EU environmental policy (see below). The second was growing pressure from national environmental pressure groups, which learned to exploit the political opportunities presented by integration and Europeanization to achieve considerably higher environmental standards than might otherwise have emerged through national action. Third, these powerful but informal thrusts of European integration were nourished and, in turn, formalized by supranational actors. In the case study chapters, there are numerous examples of the European Court of Justice (ECJ) intervening to maintain and, on occasions, raise environmental standards in the EU. Important Court rulings in the 1960s and 1970s helped to legitimize the Commission's activities and tighten the legal framework of compliance with EU rules. Finally, although as an issue the environment is much more politicized, the Commission continues to employ a 'pragmatic, incrementalist approach' (Rehbinder and Stewart, 1985, p.246; Héritier, 1999, uses the word 'subterfuge') to expand the environmental *acquis*. The case study chapters show how it develops networks of expertise to build consensus around certain proposals, exploits legal rules to push legislation through faster (the 'Treaty-base game': Hovden, 2002), insulates drafts from external scrutiny and finds legal hooks to harness environmental initiatives to single market legislation. Many of these devices were used to spectacular effect in the period before the SEA (Chapter 4).

Maturity

In spite of these very uncertain beginnings, the environment has evolved into a mature area of EU activity. The environmental *acquis* now comprises well over 500 legislative items and, until recently, was one of the fastest growing areas of EU activity. The SEA was particularly important because it helped to entrench and formalize the *acquis* that had developed informally, by giving it a firm legal footing. Crucially, the SEA provided the institutional means to achieve still higher standards by altering the decision rule in the Council of Ministers (CoM) to qualified majority voting (QMV) for proposals linked to the single market. It also promoted greater intersectoral integration via the environmental policy integration principle (EPI; this is the idea that all policy sectors should take steps to protect the environment), and emboldened the Commission to venture into new areas such as freedom of information and eco-auditing that would have been extremely difficult to justify under the Treaty of Rome. These changes were consolidated by the 1993 Maastricht and the 1999 Amsterdam Treaties (Chapters 5 and 6 respectively), which introduced the cooperation and then the co-decision-making procedures. As the most environmentally ambitious EU institution, the Parliament has used these new institutional rules to strengthen EU environmental policy, especially in the period since Maastricht.

European Union environmental policy and Britain

Even though implementation remains a persistent problem in the EU, and states continue to differ markedly in terms of their national approaches to environmental protection, the environmental *acquis* almost certainly *has* Europeanized national practices, tools and policy paradigms (Weale *et al.*, 2000). If we turn to Britain, two features of its political system have strongly conditioned its response to the development of EU environmental policy. First, the national political-administrative system has never committed itself wholly to European integration. With a few notable exceptions Britain has tended to regard the EU as a 'disagreeable necessity rather than a positive benefit' (Gowland and Turner, 2000, p.5). Whitehall was, and remains, a microcosm of these wider attitudes and beliefs. So, although Heath was a committed European, 'opinions [in 1973] were divided and "neutralism" towards the Community was sometimes the most positive of attitudes in some departments' (Willis, 1982, p.25). In these circumstances, neither officials nor their political masters could admit to being dictated to by Europe; it was better to deny Europe mattered, or nip any Europeanization firmly in the bud. However, Britain does have a reputation for administrative excellence in the EU: it prepares carefully for meetings, has a 'Rolls-Royce' mechanism for coordinating European affairs across Whitehall (see Chapter 3), and believes it has a good record of implementing EU legislation (Wallace, 1995; 1997), including environmental policy (Weale *et al.*, 2000, p.320).

   The second feature is the enduringly low political status of environmental issues in British politics. Although there have been periodic bursts of public support since 1970 (see below), Whitehall has never felt compelled to support environmental arguments. Robin Sharp (1998, p.55), the former Head of the EPG's international division, has cogently remarked that '[t]here are many environmental assumptions, right or wrong, that have to be argued within [Whitehall] that simply appear to be self-evident in the German or Dutch context'.

These two features make the deep Europeanization of British environmental policy since 1970 all the more puzzling. After all, if Britain has generally been suspicious of European governance and decidedly modest in its environmental aspirations, how are we to explain the deep-seated and wide-ranging Europeanization of British environmental policy since 1970? This book tries to answer that question by investigating the interdepartmental politics surrounding a series of 'big' and 'small' decisions (see Chapter 1). It has been conjectured that the distinction between 'big' and 'small' decisions is particularly important in relation to a semi-detached Member State such as Britain. Britain has, it is claimed, a reputation for contributing constructively to the processes of daily policy-making, which belies its much more well-known reputation for behaving 'awkwardly' in the 'big' set-piece events such as the negotiation of the Maastricht Treaty (Aspinwall, 2000). We will return to this point in the concluding chapter.

Europe and the machinery of British government

This book examines the co-evolution of integration and Europeanization from a *departmental* perspective. Departments have been aptly described as the 'nodal point' of the British administrative system (M. Smith, Marsh and Richards, 1993, p.569), and the 'terrain on which policy is made' (M. Smith, Richards and Marsh, 2000, p.146). Therefore, how they respond to the processes of European integration should interest scholars of national *and* EU politics.

Departments and 'departmentalism'

The vertical divisions between different departments are one of the central features of British administration and an enduring source of political conflict. According to a former civil servant:

> Much of the work of Whitehall is institutionalised conflict between the competing interests in different departments. Each department will defend its own position and resist a line that, while it might be beneficial to the government as a whole or in the wider public interest, would work against the interest of the department.
>
> (Ponting, 1986, p.102)

It is a commonplace that each department has its own departmental culture, which is deeply rooted in history (M. Smith, 1999, p.130). In fact, the first thing newly-created departments do is search for a coherent 'view' of the world and a distinctive departmental 'culture' of working with other actors.

Europe adds a new and interesting dimension to the interdepartmental conflicts in Whitehall, because although an individual department stands at the nexus between EU and national policy, under the well-established Cabinet principle of collective responsibility, it is also responsible for representing the *collective* (or governmental) position of *all* Whitehall departments in Europe. More often than not, departments disagree on what that common 'interest' should be and internal conflict ensues (Chapter 3). And yet the Rolls-Royce coordination mechanisms usually achieve a working consensus which allows departments to 'sing from the same hymn sheet' in Europe. The coordination process is meant to internalize any conflict within Whitehall, but the EU has nonetheless created a new political system in which interdepartmental conflicts are continuously played out. Thus, each department has an obvious incentive to ensure that its departmental interest is reflected in the national interest communicated to Brussels. At the same time, every department will also want to know how cognate departments are reconciling the pressures created by European integration and Europeanization in their own policy domains. In effect, the multi-levelled environmental politics described above also extend horizontally across Whitehall. The next chapter reveals that our two theoretical approaches adopt a sharply contrasting view of the vertical and horizontal relationships which arise when central government departments engage in the processes of integration and Europeanization.

The Europeanization of national departmental life

Europe reaches deep into departmental life because EU law takes precedence over British law and creates justiciable obligations. How have departments responded to this challenge? The general pattern is for the most Europeanized departments to devolve European work as much as possible to line divisions where it is completely mainstreamed with 'domestic' policy, whereas the less affected departments centralize European work in European coordinating units. The DTI established a European unit in 1971, but today, the trade side of the department is almost completely centred on Europe. As early as 1973, MAFF had two divisions dealing with EEC and, by the early 1980s, Europe was '*the* dominating premise' of every official and Minister's professional life (H. Young and Sloman, 1982, p.73).

The DoE, on the other hand, was not under immediate pressure to 'think European' because the scope of EU activity was still fairly limited in the early 1970s. Unlike MAFF and agriculture, there was virtually no environmental *acquis* for the DoE to adopt in 1972–3, and it was only very tangentially involved in the entry negotiations (Sharp, 1998, p.33; Hannay, 2000). The DoE was, of course, only created in 1970 as one of Prime Minister Heath's sprawling 'super-ministries' which, at its founding, employed over 74,000 staff (Draper, 1977, p.9). It brought together three very different departments, namely the Ministry of Public Buildings and Works (which provided office accommodation and other civil service buildings), the Ministry of Housing and Local Government, or MHLG (which controlled local authority financing), and the Department of Transport (DoT), which contained a mixture of executive and supervisory functions. Crucially, these three departments addressed local issues such as urban quality, transport and sanitation. None of them was 'environmental' in the modern sense of the word, although the MHLG took the lead in dealing with technical issues such as air and water pollution. Finally, the Central Unit for Environmental Pollution (CUEP) was parachuted in from the Cabinet Office to coordinate the DoE's new 'environmental' (i.e., pollution) functions. The CUEP was supposed to be the DoE's new 'environmental conscience', and interface with international environmental bodies.

Like most other weakly Europeanized departments, the DoE centralized its European work in the international part of CUEP (later re-christened the CDEP, or Central Directorate for Environmental Protection). Interestingly, the CUEP was a new and (to most DoE civil servants) somewhat alien body. Only created as a separate Cabinet Office unit in 1969, it was moved to the DoE in 1970. The EPG did not create a separate European coordinating unit (EPEUR) until 1990, by which time literally hundreds of environmental EU regulations had been adopted. Before then, European issues were subsumed into the work of the CUEP's international division, EPINT (Environmental Protection (International)), which interfaced with the UN and other international bodies. Throughout the 1970s and 1980s, European environmental work resided within EPINT, which gained a reputation for being a specialized policy 'ghetto'. EU environmental work touched only a very small fraction of EPG's staff and virtually no one else in the other, 'non' environmental groups within DoE. To put this into context, by the early 1990s all Grade 7s (i.e., middle level managers) in MAFF had worked on EU business by the time they attained that grade. In the DoE in 1993, 'there [was still] no such expectation, and it is quite possible that officials at senior levels will have [had] no direct experience of [EU] work' (DoE, 1993c, p.16; and see Table 2.2).

Table 2.2   Total number of Whitehall staff working on EU policy, 1989

Department	Total number
MAFF	300–400
DTI	92
DoE	60
Foreign Office	56
Employment	33
Energy	12
Lord Chancellor's Office	12
Health	8
Defence	0

Note: Bear in mind that DoE and DTI were considerably larger than MAFF at this time.
Source: Based on Department's answers to a series of Parliamentary questions (HC. Hansard (Written Answers), Vol. 149, 6 March 1989, col. 338; 15 March 1989, cols 215–57; 21 March 1989, col. 495).

The Department of the Environment: a department in search of a departmental culture?

In theory, the DoE should have embraced European rule making (see Chapter 1), but it inherited a departmental culture from its predecessors that was neither 'environmental' (in the modern sense of the word) nor 'European'. Heath maintains that he did not create the DoE for environmental reasons (Heath, 1998, p.314). However, the very first Secretary of State for the Environment (SoSE), Peter Walker, realized that his new department needed a unifying culture and seized on the environment (Radcliffe, 1985, p.208). However, Walker's successors (see Table 2.3) struggled to maintain his integrated environmental approach to management, and by the mid-1970s the various wings of the department were said to be operating in a 'semi-autonomous fashion' (Radcliffe, 1991, pp.113–15). As the British economy slipped into recession and the political demand for environmental measures waned, Labour SoSEs such as Crosland and Shore concentrated on the more economic parts of the DoE's vast empire, and issues such as pollution control and habitat conservation took a back seat. Insofar as there was an operational 'environmental policy' throughout the 1970s, it was overseen by professional agencies such as the Industrial Air Pollution Inspectorate (IAPI) rather than the DoE, following the well-established policy style of devolving operational decisions to local agencies (Jordan, 1998b). Throughout the 1970s, EPG remained somewhat of a backwater in the Department, employing just 3.2 per cent of the DoE's staff in 1980 (McQuail, 1994). Tony Fairclough (1999), a former Head of the CUEP, admits that the DoE 'certainly wasn't a department of the environment as such, and it wasn't even a department for the environment, which a lot of people criticised it for. But it was inevitable in a way because it was a construct from three totally different ministries and totally different cultures.'

During the formative years, key figures such as Peter Walker and Martin Holdgate tried to use international bodies such as the United Nations to raise the profile of environmental issues in the department (Lowe and Ward, 1998, pp.9–10). However, Holdgate (2000) remembers that it was hard going: 'nobody in the DoE wanted to go international. There was a huge domestic inward lookingness. They didn't want to waste time in international meetings. I was let loose to negotiate with minimal involvement from the other sectors of the department.'

As an organization, the DoE shunned the opportunities presented by the EU to strengthen national environmental policy by uploading concepts to Brussels. This was not so much because Ministers were *anti*-European, but because they were generally uninterested in the environment as a political issue. It is remarkable how few SoSEs were personally committed to environmental protection. Of those that passed through the department in the 1980s (and a lot did; see Table 2.3) *en route* to better postings, most concentrated on reforming local government finance (e.g., Baker, Heseltine) and tackling inner-city decay (Shore, Heseltine), which were issues that dominated British political agendas but did not directly involve a strong EU dimension.

One of the inevitable casualties was *European* environmental work. By the mid-1980s, this was so politically downgraded that the DoE

Table 2.3   Secretaries of State for the Environment, 1970–2000

Name	Year	Party
Walker	1970	Conservative
Rippon	1973	Conservative
Crosland	1974	Labour
Shore	1976	Labour
Heseltine	1979	Conservative
King	1983	Conservative
Jenkins	1983	Conservative
Baker	1983	Conservative
Ridley	1986	Conservative
Patten	1989	Conservative
Heseltine	1990	Conservative
Howard	1992	Conservative
Gummer	1993	Conservative
Prescott	1997	Labour

regularly sent a Parliamentary Under-Secretary (i.e., not even a Minister), William Waldegrave, and fairly junior civil servants to Environment Councils to negotiate important Directives (Sharp, 1998, p.34). The best and brightest civil servants gravitated towards the more politically salient areas, such as housing and local government. Holdgate (2000) remembers that:

> [t]here were one or two people who did go off and work in Brussels and in the OECD, but it was not considered a high priority. Indeed, there were some senior officials who would advise promising people that it would be detrimental to their career – 'if you move out of the department your re-entry is very difficult' sort of line'.

The DoE had little inherited experience of European work to draw upon and in the absence of strong popular support for European integration (in 1975 the country voted in a national referendum to remain a Member State), the incentives to upload national policy were, as one former Head of the CUEP recalls, extremely weak: 'I don't recall areas where we were pressing the Commission to develop new initiatives as a result of our national priorities. Commission proposals were in a sense *distractions* from the traditional process of national policy development' (T. Hall, 1999; emphasis added).

Of course not every single individual in every division of the DoE responded quite so negatively. For instance, those supervising the regional funds (where there were obvious financial incentives to be proactive) did positively engage themselves in EU affairs, but they were an exception.

Throughout the 1970s and early 1980s there was an unstated and untested assumption within the EPG that EU environmental policy would not amount to much. Holdgate (1983, p.11) thought the Commission was simply 'on the sidelines, watching what was happening in the Member States'. Even committed Europeanists had to admit that the EU was 'mundane' in comparison to national policy 'and the wheels [do] grind excessively small' (Haigh, 1984, p.7). It is indicative that Walker prepared a lot more carefully for the 1972 Stockholm conference than the Paris Summit. A working group chaired by the Chairman of the standing UK Royal Commission on Environmental Pollution (RCEP), Lord Ashby, reported that 'effective measures . . . should if possible be national . . . international action should be reserved for international problems' (Ashby, 1972, pp.76, 79). This advice was then re-transmitted by the DoE to the Stockholm conference delegates, who were informed that supranational policy 'will be most usefully concentrated at truly global problems' (DoE, 1972), by which was meant that the EU should do very little.

It is fair to say that the EPG was deeply imbued with the same belief (or 'pride') in the quality and adequacy of British approaches which had developed among British environmental policy elites over a century or more (Hajer, 1995). On the first page of a guide entitled 'Pollution Control in Great Britain: How it Works', the CUEP claimed that Britain was 'at a comparatively advanced stage of development and adoption of environmental protection policies' (DoE, 1978a, p.1). In effect, the DoE told the Commission not to waste its time developing proposals because Britain was 'well placed to cope with its own environmental problems' (see D. Evans, 1973, p.43). According to Derek Osborn (2000), Head of the EPG (1990–5), 'people . . . felt they knew how to do pollution control . . . and they felt resentful about early Commission initiatives. So in the 1970s and 1980s they found ways to dilute them, seek derogations, go about things slowly.'

The Europeanization of British environmental policy

1970–3: the birth of modern environmental policy in Britain[1]

In comparison to other national environment departments, the DoE was extremely slow to upload legislation to Brussels (see Table 2.4). An internal note on the EU standstill agreement (a gentleman's agreement to notify the Commission of all new items of domestic legislation: see DoE, 1982b) offers one explanation: 'Notifications from the [UK] have been comparatively few and have not led the Commission to take action. The reason would appear to be the general advanced state of environmental legislation at the time of accession and its character.' As the very first industrialized country, Britain did indeed 'discover' pollution somewhat earlier than other European countries, and it did pass some of the first anti-pollution legislation anywhere in the world. Thus, when the EU began to take an interest in the environment, Britain had relatively little new policy to submit to the Commission under the standstill agreement because it had effectively already legislated.

This put the DoE at a huge 'first mover' disadvantage, because the Commission would often base its proposals on submissions made under the agreement (Rehbinder and Stewart, 1985, p.259). That said, it is debatable whether the DoE (1982b, p.2) would have been any more proactive if it had had new policies to upload. It perceived that '[s]ince

Table 2.4   Notifications under the 1972 'standstill agreement', 1973–80

	Draft legislation	Administrative measures	International agreements	Others	Total
UK	6	0	0	0	6
Germany	28	2	2	2	34
Belgium	1	3	1	0	5
Denmark	1	24	4	0	29
France	26	6	3	6	41
Ireland	2	0	0	0	2
Italy	4	0	2	0	6
The Netherlands	6	17	0	0	23
Luxembourg	1	0	0	0	1
Total	75	52	12	8	147

Source: Wurzel (1999).

accession *there has been little call on the [UK] to set the pace* [in the EU] creating new measures for environmental protection' (emphasis added).

To a large extent, the DoE's attitudes mirrored those in civil society (see above). Environmentalists on the more radical wing of the green movement cautioned against any involvement with the EU, and more mainstream groups regarded it as 'terra incognita' (von Moltke, 1983, p.37). Above all, there was a widespread belief that the EU (or any of its member states) would and should not do things that disturbed British practices:

> The need for an information agreement . . . seems self-evident if the Commission is to carry out its tasks properly. There remains a duty on national governments to respect their obligations as members of the Community not to advance proposals for legislation in isolation or which would affect the functioning of the common market or the operation of the [First] Action programme.
>
> (DoE, 1982b, p.3)

1973–9: financial restraint and political conflict

The high hopes and expectations of the DoE's formative years soon gave way as Europe entered a deep recession. In Britain, one of the first casualties was the flagship Control of Pollution Act (1974), which was never fully implemented. In 1974, EPG lost an important lever of national air pollution policy to the Health and Safety Executive (Weale *et al.*, 2000, p.224). British environmental policy was obviously in the doldrums, but the Directorate-General of the Environment (DG Environment) continued to look for opportunities to expand the *acquis*. The misfit between Britain's ambitions and those of the EU widened when, in 1974, the Commission issued ambitious proposals to reduce emissions of certain dangerous substances to water. A Directive was eventually agreed in 1976 after much wrangling. It contained extensive exemptions for Britain to reduce the misfit with national practices, but the whole episode soured British–EU relations for years to come. Andrew Semple (1999), a former Head of the Water Quality division, feels that the dispute became self-perpetuating:

> [it] went on and on and on and on . . . Ministers, without, I suspect fully understanding what the issues were, saw it as a question of 'we have a British policy and this is our British policy'. And the Treasury and the DTI saw it in terms of 'we have a British policy and it's a cheaper policy'. These issues raged on for years and years and years, without much consideration of what they were actually about. They became like parrot cries . . . It made it very difficult for us in the Department to convince other countries that we were not simply collecting sewage and putting it straight through the tap.

The negative position adopted by the DoE during the negotiations was partly a reflection of strong industry lobbying transmitted through the DTI as well as the need to protect the publicly owned water industry (which the DoE sponsored) from expensive investments. However, the 'pre-precaution' view that waste should be externalized along pipes into the sea rather than treated on land was also an integral element of the DoE's departmental culture (see Martin Holdgate's 1979 book on the principles of pollution control), especially among the many scientists who worked in the powerful water and air directorates (Jordan, 2003). In 1975, Denis Howell, a Europhile politician who negotiated the Dangerous Substances Directive, said that Britain had agreed to the First Environmental Action Programme:

> confident of our own efforts and experience in cleaning up and maintaining the environment in this country. Without being complacent we feel that that we have not allowed things to get out of hand and in the main have kept pollution under control and have plans . . . to make further improvements.
>
> (Howell, 1975)

The DoE continued to devote a great deal of time to explaining the scientific underpinnings of the 'British approach' to foreigners well into the 1980s (Waldegrave, 1985).

The relationship between Europe and Britain was not entirely one way; there were areas where the DoE successfully took the initiative in Europe, or at least supported European solutions (e.g., the Birds Directive (see Chapter 8), the 1981 Regulation on whales and the 1975 Directive on waste) but they were far 'fewer than might have been expected of a country with such a well established environmental policy' (Haigh, 1984, p.302). Why was this? Around this time, the domestic political profile of environmental work was low, so Ministers and senior civil servants chose to stay well clear of it. The long-serving Head of CUEP's International Division, Fiona McConnell (1999), believes there was never an incentive to upload policy to Brussels: 'it was very difficult to take the initiative in Europe . . . We were always reacting. You got no credit and considerable discredit for throwing a stone into the Brussels pond and looking to see what happened to the ripples. "Leave it alone" is what we were told!'

Those working in Brussels certainly saw very few British ideas coming their way. The then Chairman of the European Parliament's Environment Committee, Ken Collins (2000), recalls that: 'European policy was seen as an extraneous imposition to be resisted. The DoE . . . like most other departments of state . . . understood Europe only in intergovernmental terms . . . They saw Europe as a brake on British policy even when British policy wasn't actually moving!'

According to a former UKREP official:

> our job in Brussels was to kill proposals before they took root, to neuter those that did come up for active discussion and to find ways of reducing the practical effect of those that did get agreed. But like Canute, in the end it was an impossible job because we were always rowing against the political tide in Europe.

To take one example, in 1980, the Water Directorate's objectives were minimal and entirely negative: '[t]o gain acceptance of the UK policy of control . . . ; to avoid proposals involving . . . . industry in unwarranted expense; to ensure that Parliamentary scrutiny committees are kept informed' (DoE, 1980b, Part 3, Water Directorate, p.22).

1979–88: the Europeanization of national policy

Under Mrs Thatcher's premiership, a succession of Secretaries of State for the Environment devoted their time to reducing the size of the Department, streamlining the planning system and making regulation less onerous. The DoE's major goals in this period were improving inner cities, reforming local government finance (culminating in the infamous poll tax) and privatizing state-owned assets such as water. During the 1980s, the DoE was hit particularly hard by cuts in the civil service, total numbers falling by nearly 50 per cent in the period 1976–91 (Hood, 1995, p.114), but the EPG fared particularly badly, as revealed by Tables 2.5 and 2.6. The CUEP, the EPG's main line of communication with the EU, lost 30 per cent of its staff between 1980 and 1985, which was twice the average across the whole of the civil service (Environmental Data Services Ltd., 115, pp.4–5). Staff were withdrawn from international bodies and inter-departmental committees, and production of major policy reviews ceased. In 1983, it was reported that civil servants were 'largely preoccupied with routine administration and increasingly with reacting to international initiatives rather than evolving the UK's own environmental policies' (ENDS, 107, p.12).

By contrast, during this period, EU environmental policy was at its most active and innovative. The DoE tried to close the growing misfit between national and EU policy employing the national veto, typified by the five-year battle against the EIA Directive even though it merely formalized existing land-use planning practice in Britain (see Chapter 10). Every effort was also made to neuter legislation that had already been downloaded from Brussels, typified by the identification of just 27 bathing beaches under the Bathing Water Directive (Chapter 7).

Waldegrave was probably the first Minister who actively championed the environment and genuinely understood the importance of the

Table 2.5   The relative size of the EPG (1980–93)

Year	Number of staff in the EPG	% of total DoE staff
1980	333	3.2
1981	260	2.6
1982	265	3.0
1983	250	3.2
1984	248	3.8
1985	252	3.9
1986	288	4.4
1987	370	5.8
1988	418	6.4
1989	528	8.0
1990	522	10.3
1991	555	10.6
1992	804	17.9
1993	827	18.4

Source: McQuail (1994), p.52.

36

Table 2.6 Staffing levels in the EPG (1980–9)

Section	1980	1981	1982	1983	1984	1985	1986	1987	1988	1989	% change (1980–9)
CUEP	74.5	68	63	58	73	75	89	86	63	65	–13
Air, noise and waste	96.5	94	95	95	77.5	88	96	113	45	55	–43
Water	100	98	99	90	92	82	85	94	69	100	0
Water privatization							27	27	64	73	n/a
Economics, science and statistics	22	22	14	9	9	8	10	9.5	9.5	9.5	–57
Radioactive waste	12	12	11	12	15	18	21	21	19	22.5	+87.5
HMIP							69	75	214	190.5	n/a

Notes: Water privatization and the creation of Her Majesty's Inspectorate of Pollution (HMIP) in 1987 greatly increased the size of EPG relative to other parts of the DoE (see Table 2.5). EPG's international arm, CUEP, fared particularly badly, as did the DoE's scientific support functions.
Source: MINIS reports, 1980–9.

European dimension. He recognized that European integration was about more than just 'cherry picking' Directives because they happened to 'fit' national needs. He tried to upload ideas to Brussels, but lacked the political clout consistently to win important intra- and interdepartmental battles. Over-defensive in its attitude to the EU, shorn of financial resources and neglected politically, EPG did not play the game of EU politics as well as it could have: 'We were not good at negotiating in Europe . . . Every issue had to stand on its own. We fought battles on each individual issue, but we failed to realise . . . that Europe was an endless game of give and take, and we never gave and took' (Semple, 1999).

On the surface, British attempts to stymie the Europeanization of policy by saying 'no' in the Environment Council appeared to be working well. But as the environmental *acquis* 'mutated' (Weiler, 1991; see above), EU policy began slowly to insinuate itself into Britain, forcing the DoE to revise its view of EU Directives. In particular, the Commission started to oppose the DoE's attempts to close misfits by subverting Directives at the implementation stage. In fact, the then environment Commissioner (and former Labour MP), Stanley Clinton-Davis, seized every opportunity to extend any misfits which appeared by sanctioning infringement proceedings against Britain when it simply breached the spirit of EU law, especially in the water sector (Chapter 7). These culminated in judgments by the ECJ, which sharpened the legal force of EU Directives. National environmental groups such as Greenpeace, which opened a Brussels office in 1988, began to adjust their campaigning activities to exploit these new political opportunities. Interestingly, both Greenpeace and Friends of the Earth (FoE) adopted European terms and points of reference (*'the* Dirty Man of Europe': Rose, 1990) to attack British departments. As the political centre of gravity shifted upwards to Europe, other, previously sceptical, national bodies began to side with the EU. For example, for the first time in 1981 the respected House of Lords select committee on the EU backed the Commission against the DoE on the issue of EIA (Chapter 10). In 1984, the RCEP, no less, maintained that Britain's performance in Europe:

> has not always been as strong as it might have been, and . . . on occasion . . . may have harmed [its] international standing. We see a need for a more positive attitude and a willingness to seize the initiative . . . not merely to ensure that national interests are best served but *also to give a lead internationally on best possible practice in tackling pollution.*
>
> (RCEP, 1984, para. 1.24, emphasis added)

By the mid-1980s, Britain was markedly out of step with the rest of Europe, weighed down by a very poor environmental reputation. The then Head of the EPG, Derek Osborn (1997, p.5), offered the following diagnosis: 'We were failing to attend properly to emerging problems and we were not investing sufficiently in solutions. We compounded the problems by using disingenuous arguments to defend our position in some cases . . . [and used] . . . the absence of 100% scientific proof . . . [as] a lame excuse for inaction.'

Eventually, Britain suffered the fate of those states that persistently download policy from the EU, as it was hit by a succession of political crises. In Britain's case, the crises arose when EU policies began to escape from the narrow confines of the EPG and intrude into the most import-ant areas of the DoE's business, such as water and then energy privatiza-tion (Chapter 7). Until then, the Europeanization of British policy had occurred slowly and imperceptibly, at a 'subterranean' level in the department (see above). Many of the early Directives were negotiated by scientists and technical experts in EPG. Lawyers and more senior civil servants tended not get involved. But in the mid-1980s, the Department was forced to reassess its view of EU environmental policy as matters of 'low' politics suddenly became matters of 'high' politics. It was a rude awakening for the DoE's high command, which began fully to realize (to its great surprise and political cost) that it could not achieve policy objectives in 'non' environmental areas, because of the seemingly 'tech-nical' commitments that EPG had (perhaps too easily) entered into: 'We gradually began to realise there were elephant traps under everything. We didn't always understand the complications, but we were aware they [Directives] were very tricky and influential. Ministers . . . simply didn't understand what was going to be involved in implementing them' (Semple, 1999).

The sense of crisis became so acute that, eventually, the Prime Minis-ter, Margaret Thatcher, was forced to intervene. She ordered a major interdepartmental review of environmental policy (ENDS, 123, p.3). Among a raft of possible measures, it identified Europe as a major prior-ity, claiming that the DoE 'did not seem to have the capacity to keep up with outside and international developments, let alone keep ahead of the game' (ENDS, 133, p.3). The sharpening conflicts with Europe also forced the then SoSE, Nicholas Ridley (1986–9), to involve himself much more fully in EPG's affairs than his predecessors. With his active support, EPG started to win important interdepartmental battles against the DTI, MAFF and the FCO on issues that had a very strong EU dimension, such as acid rain (1986), North Sea pollution (1987) and ozone depletion (1987). These supranational activities in turn demanded extra resources, and EPG began to grow in size and political stature relative to other parts of the DoE (see Table 2.5). The party politicization of the environment accelerated when Mrs Thatcher made a headline speech to the Royal Society in 1988. The following year, the then Head of the CUEP, John Hobson, recorded that: 'On all fronts we have succeeded in improving the UK's international image; partly by removing some long running sores, partly by going onto the offensive in some areas; and *partly by getting better at saying no gracefully'* (DoE, 1989, p.9).

Hobson's comments indicate that important but subtle changes were occurring inside the DoE as it sought to reach out to, and communicate with, the EU. Among a number of innovations, in 1988 CUEP organized an unprecedented meeting with the incoming Presidency of the Environment Council and a Ministerial trip to the EP (DoE, 1989, CDEP, 2.04, p.12). It also commissioned Nigel Haigh's Institute of European Environmental Policy to report on the implications of the SEA. Haigh concluded that:

A government department . . . structured to deal with national problems and to developing and implementing national policies must now adjust [itself] to being part of a larger system that exists to develop and implement EC policies . . . This implies that the department will need to develop sources of information about environmental problems throughout the [EU], about the policies of other Member States, about the institutional structures for handling policy within the Member States and about the effectiveness of implementation of these policies.

(Haigh and Baldock, 1989, pp.47–8)

The IEEP also calculated that 300 EU environmental laws had accumulated since the early 1970s. The official in EPINT who coordinated EPG's input to the Environment Council at this time said the figure of 300 raised eyebrows:

We were novices almost. Here was a slightly academic group saying you really must take Europe a little more seriously . . . It gave us evidence to draw to the attention of Ministers that this was an important area of work . . . There was a . . . vague impression that a lot was happening in Europe but nobody had actually quantified it.

(Shaw, 2000)

1988–2000: the Europeanization of the Department of the Environment

In July 1989, Thatcher replaced Ridley with Chris Patten, a younger and more media-friendly politician. A committed Europhile, he nailed his colours firmly to the mast by attending the Environment Council in person. Patten also reacted to the rising curve of international and European work by creating EPEUR within EPG, headed by a former UKREP counsellor, John Plowman. Plowman (2000) believes Patten was the first SoSE to appreciate the need to 'to clean up [the UK's] act by developing national environmental policies that *meshed in* with European ones' (emphasis added). Patten demanded a more proactive attitude to European affairs *and* a willingness to respond positively to, rather than to deny, the Europeanization that had already taken place. His Minister, David Trippier (1991, p.9), reminded the Department of the need to be more self-sufficient in the EU, using terms first employed by Heath:

> it is vital that we play a leading role in the formulation of new legislation. To do this we have to be engaged. We have to be in constant dialogue with the [Commission] and other member states. We have to be constructive and proactive in our approach. We cannot be influential if all we do is barrack from the sidelines. *We have to learn to think European.*
>
> (emphasis added)

Two years later in 1993, civil servants duly reported that:

> Our aim has been as far as possible to take a proactive approach. We have a rolling programme of discussions at least annually at both official and Ministerial level with many Member States, and we encourage policy Divisions to establish direct contacts with their opposite numbers. For the future we need to improve links with the European Parliament . . . so we have strengthened briefing arrangements and put in place a programme of informal contacts.
>
> (DoE, 1993b, Directorate of Environment Policy and Analysis, p.4)

These internal changes were partly driven by the need to come to terms with the Europeanization of British environmental politics and policy, but they were also lagged responses to alterations in the political geometry of the EU, namely the arrival of new, Mediterranean Member

States with weak environmental policies and the unremitting extension of QMV (see above). These two changes altered the DoE's self-perception: first, it was no longer the environmental laggard (other countries now had worse records); and second, it could not take refuge behind the national veto. In a sense, the Department began to realize that it was in its own *self-interest* to engage with Europe by seeking new allies and uploading its own policies, rather than downloading them from other countries. As one former Head of the Water Directorate remarked '[i]f you are just reactive you end up with a bad image . . . being proactive is the best form of defence' (Summerton, 1999).

From the Head of EPG downwards, the Department tried to plug into Europe more effectively. Using a mixture of formal and informal channels of influence, it set about uploading homespun policies to the EU, such as Integrated Pollution Control, or IPC (Chapter 9), eco-auditing and environmental management. A new cadre of top officials took a strategic decision to use Europe to advance EPG's new-found 'environmental' objectives:

> [W]e were trying to shift the department's position in European environmental negotiations from a 'no, because' position to a 'yes, but' position. So often we were in a defensive position because our style was to say, 'no, because x, y, z about the proposal is wrong' . . . Stance is all important in the European context.
>
> (Summerton, 1999)

These changes were cemented into place by the next SoSE, John Gummer. He felt the DoE was still not a wholly *European* department:

> They [civil servants] thought of it as something over there. There was a sort of dissatisfaction that things to which they had signed up to were now turning out to demand much more than they expected. You have to get yourself into the mood of realising that this is your responsibility in the UK. You signed up to it and it's not something you can shovel off to politicians in Brussels. Until you do that you don't actually do your job properly. The Department hadn't quite learnt that it had to take responsibility for what it had agreed. It was a question entirely of coming to terms with reality.
>
> (Gummer, 2000)

Gummer's 'European professionalism' initiative (as it became known within the EPG) had a number of key elements, including: a renewed effort to place DoE officials in UKREP; a more structured programme of training and awareness raising; closer contacts with the Commission and other national environmental departments; and the publication of a handbook on how to negotiate successfully in Europe (Humphreys, 1996). Gummer regards the publication of the handbook as a pivotal moment: 'it exemplifie[d] a department that had become European. You could not have written that book if you weren't part of a department whose whole atmosphere and attitude was European. Imagine someone proposing that you wrote such a book in 1983. No, it wouldn't have happened' (Gummer, 2000).

Finally, Europeanization and then European professionalism have encouraged the DoE to reflect critically on its reputation in Europe. An internal Ministerial report produced in 1996 appeared to confirm what people such as Nigel Haigh had been trying to tell the Department since the 1980s, namely that Britain had a poor reputation, which had undermined the DoE's negotiating capital in Europe. The report recommended: better public relations to explain Britain's strengths and weaknesses; different negotiating tactics (e.g., adopting fewer hard-line positions in the Environment Council and dealing with proposals from other actors in a more open-minded way); a stronger input from the DOE's lawyers; and a more reflexive attitude to European affairs (i.e., thinking about how policy statements are perceived by other actors, operating in different circumstances). It is a measure of how far the department (and the EU) has travelled since the 1970s that these issues are now being actively considered at the very highest levels in the Department.

Conclusion

At its founding in 1970, the DoE was the first and by far the largest environment department in the world, with sufficient breadth of coverage to be an effective voice in Whitehall. With the DoE and a SoSE who took an active interest in environmental politics, Britain could legitimately claim to have been a European environmental pioneer in the period 1970–2 when EU environmental policy was in its formative stages. However, the DoE inherited an attitude of mind from its predecessors that was neither 'environmental' nor 'European'. Under the direction of the Chief Scientist, Martin Holdgate, the CUEP invested more time in international environmental diplomacy than the EU, which appeared intent on discussing only the most narrow and trivial of issues that had little direct relevance to Britain (a country, remember, which thought it had invented modern environmental policy). Politically speaking, Britain underwent the transition to ecological modernization (Weale, 1992) slightly later than countries such as The Netherlands, Denmark and Germany. For a long time, the DoE struggled in Europe because domestic political and economic support for environmental protection was out of kilter with events in the 'leader' countries and DG Environment's desire to adopt ambitious environmental policies.

However, domestic politics alone cannot account for the DoE's insular and reactive attitude to the EU. Deep down, the DoE was a department of regional planning and local government; matters which, even today, are only very weakly Europeanized. The fact also that members of the scientific and legal elite in Britain, such as Eric Ashby, were so deeply critical of the Commission's early interventions easily convinced the DoE's high command that the EU should be watched, but not actively steered. Ashby, for example, believed that some of the early Directives were 'a repudiation of the lessons we have learned from 160 years of our own history' (Ashby and Anderson, 1981). Their Lordships (House of Lords Select Committee on the European Communities, or HOLSCEC, 1978, p.8) even questioned the *vires* of the *acquis*.

The 1980s witnessed a massive revival in EPG's fortunes as environment entered the political mainstream in Britain. Crucially, however, the politicization of environmental issues in Britain was driven, in large part, by EU pressures, specifically those arising from the failure to 'fit' Europeanization and integration. The 'Dirty Man' tag was really just a crude label for a host of organizational, political and financial problems that afflicted UK environmental policy in the 1970s and 1980s. The DoE was undeniably a victim of these external pressures, but it did very little itself to try to resolve them. In the 1990s, the Department tried to regain control of environmental policy-making, though often in response to the gathering pace of events at the European level rather than in anticipation of them.

Having examined the Europeanization of British environmental policy in very broad terms, we now turn to our two main theoretical perspectives, before documenting empirically the dynamics of European integration and Europeanization in the seven case study areas.

3
Theories of European Integration and Europeanization

Many different theories touch upon the role of departments as mediators (or 'fitters') of European integration and Europeanization. This chapter examines two of them in some detail: state-centric and process-based. On closer inspection, both have a lot to say about 'states', integration and Europeanization, but they are relatively silent on the precise role played by individual departments. Indeed, one approach, the state-centric, subsumes departments into 'the state' and leaves it at that. On this view, a departmental perspective offers little because departments *per se* do not matter. Process-based theories, on the other hand, argue that the institutional context of political action and historical processes *matter* (i.e., when and where decisions are taken in the EU, and which parts of the state take them, significantly affects policy outcomes). This chapter examines how the two approaches theorize: (1) the origins and articulation of 'state' preferences; (2) the role of departments as either facilitators or opponents of integration Europeanization; (3) what motivates states/departments to pool national sovereignty during the integration process; (4) the ability of states/departments to manage the domestic impacts of the integration (i.e., Europeanization) that they have knowingly or mistakenly sanctioned.

State-centric theories

The 'holy grail' of integration theory since the 1950s has been to establish who or what steers the process of European integration. Do the Member States, acting either individually or collectively, exercise strong leadership, or do non-state actors play the decisive role? Theories of the EU offer very different responses to this question.

Fundamental premises

Moravcsik's state-centric or liberal inter-governmental (LI) theory assumes that, throughout the history of the EU, political leaders have pursued integration in order to secure their national economic interests. Seven of its guiding assumptions are noteworthy. First, the *formal* bargaining that occurs between states in the European Council and the various sectoral formations of the Council of Ministers more or less determines the course of integration. Second, states are rational, self-interested actors that assimilate national societal demands at the domestic level then negotiate with other states at the European level to achieve the best possible outcome. Third, domestic groups seek to impose constraints on the state to ensure it acts in their own interests. However, domestic political systems are biased against diffuse political interests, who find it harder to mobilize into groups to articulate their preferences. In supporting diffuse public goods such as environmental quality that benefit large numbers of individuals by a relatively small amount, environmental pressure groups are in an especially disadvantaged position relative to economic producers. This is because the costs of environmental protection tend to be 'lumpy' (i.e., they are incurred by a relatively discrete and geographically concentrated group of polluters). Being more limited in number, industrial interests generally find it easier to mobilize into cohesive and powerful pressure groups. Moravcsik (1993, pp.487–8) writes:

> [at] one extreme, where the net costs and benefits of alternative policies are certain, significant and risky, individual citizens and firms have a strong incentive to mobilise politically. In such circumstances, unidirectional pressure from cohesive groups . . . imposes a strict constraint on government policy . . . [A]t the other extreme, where the net costs and benefits of alternative policies are diffuse, ambiguous or insignificant, and the risk is low, the societal constraints on governments are looser . . . Under such circumstances, leading politicians enjoy a wider range of de facto choice in negotiating strategies and positions.

Fourth, a state's core task is to aggregate these national preferences (the 'liberal' in LI) and then take them to the international level, where the required policies are supplied through interstate negotiation (the 'inter-governmentalism' in LI).

Fifth, states 'keep the gate' between national and international politics. In spite of the explosion of supranational lobbying in Brussels, states are the principal conduit between national interests and supranational bargaining fora. By implication, they more or less determine the extent to which integration 'fits' with Europeanization. Although they may, on occasions, appear to produce 'misfits' by acting autonomously (e.g., enforcing compliance with downloaded policies), supranational agents are generally subservient to the interests of states. Sixth, states are generally unable to make concessions beyond their own domestically determined preferences: this drives EU agreements towards the lowest common denominator of state preferences. Consequently, unintended 'misfits' are likely to be very few in number.

Finally, LI was initially designed to function as a general theory of European integration, but in more recent work Moravcsik has begun to address its impact at the national level (i.e., Europeanization). This is an important departure, because we can only really know if states secure their objectives by examining the extent to which individual policies generate the domestic outcomes that states expected. Moravcsik's implicit (and largely untested) assumption is that because states are *a priori* in control of integration, Europeanization has to follow the path selected by the core executive during the negotiation process. This includes impacts that appear to 'transform' the state (Risse-Kappen, 1996) or empower subnational actors by creating a more 'multilevel governance' structure (Marks, 1993). Moravcsik (1994) emphatically denies that these were unintended outcomes, or that Europeanization creates unintended 'misfits', reduces or undermines the coherence of the state, or mistakenly empowers non-state actors. On the contrary, because states are the exclusive representative of the nation state in the EU, they are uniquely capable of agreeing EU policies and presenting them to domestic groups on a 'take it or leave it' basis. States may even deliberately use integration to strengthen themselves at the expense of other domestic groups (i.e., by deliberately creating 'misfits' as part of a 'two-level game' to exert power over sub-national actors). States might do this as a means to push through policies (i.e., either by uploading or by downloading) that would otherwise have been blocked by societal interests (what Moravcsik terms 'slack cutting'). In one paper, Moravcsik develops this concept by arguing that 'executive cartels' of national leaders may decide to pool state sovereignty for no obvious benefit other than to simply enhance their own domestic political autonomy (1994, p.3). There is, as we shall see, very little evidence of systematic slack cutting in relation to British environmental policy (see below), but Rehbinder and Stewart (1985, p.332) argue that the German federal authorities have used the EU in the past as a 'back door' method of overcoming national parliamentary and Länder opposition to stronger environmental rules (see also Wurzel, 2002).

The 'big bangs' of European integration

In summary, state-based theorists imply that states are not only in control of integration and Europeanization, but largely unaffected by them: 'The [EU] has developed through a series of celebrated intergovernmental bargains, each of which set the agenda for an *intervening period of consolidation*. The most fundamental task facing a theoretical account of European integration is to explain these bargains' (Moravcsik, 1993, p.473; emphasis added).

Integration is said to proceed as a 'sequence of irregular big bangs' (Moravcsik, 1998, p.4), rather than cumulatively and irreversibly through a series of much smaller, but self-reinforcing feedbacks. As originally conceived, LI displayed little interest in day-to-day policy development. Individual policy sectors were simply *assumed* to develop within the terms of the big bangs, (i.e., the periodic IGCs: see Chapter 1). IGCs are convened when a majority of states decide to amend the founding Treaties of the EU. They are steered by foreign ministers, but usually small groups of very senior civil servants (normally from national foreign ministries such as the FCO) thrash out the technical details. Treaty revisions are large, set-piece events dominated by states. They are organized in accordance with the 'unicity' principle that all discussions are strictly subservient to the Heads of State meeting in the European Council. According to a participant in the Amsterdam Treaty IGC, '[w]hat mattered . . . was not what, say, Environment Ministers said to each other [in the Environment Council], but rather how each Environment Ministry influenced the position taken by his own delegation within the negotiations themselves' (McDonagh, 1998, p.208). In other words, IGCs are *precisely* where we would expect to find states acting rationally and coherently.

If all this were true, the dominant role played by the British core executive during IGCs, underpinned by the 'unicity' principle, should leave very little scope for individual departments to make 'home runs' in Europe (Weale *et al.*, 2000, p.100). Home runs occur when individual departments (i.e., not 'the state') use Europe to realize their sectoral or departmental interests (cf. the cross-government or 'national' interest) against the wishes of cognate domestic departments and the core executive. In Chapter 1 we saw that some scholars regard 'home running' as an important cause of the continuing expansion of the environmental *acquis*, whereas others emphatically deny that 'the state' is anything other than a rational, internally coherent actor. These theoretical predictions are subjected to detailed empirical scrutiny in Chapters 4, 5 and 6 respectively.

Britain: a rational actor?

Britain provides a particularly good empirical 'case' to test the claims of LI. Of all the Member States bar France, the British government probably bears the closest resemblance to the unitary, rationally acting, hierarchically structured agent found in state-centric accounts. The basic principle of British government is that, wherever possible, business should be handled at the lowest possible level to avoid overloading the system. Thus, the core executive intervenes only when matters cannot be settled at lower levels, or where there are serious implications for other departments. Membership of the EU has forced the core executive to develop extensive machinery for coordinating European business, to ensure that the British government speaks with a single, coherent voice in Europe (Edwards, 1992; Spence, 1993). These mechanisms, the more formal of which are centred on the Cabinet and the FCO, are strong and very well developed. However, as with national policy, responsibility for the detailed, day-to-day aspects of European policy-making and negotiation is devolved to individual departments, who are expected to 'think (and act) European' for themselves (see Chapter 2).

The coordination process places a duty on departments to inform one another of what they are doing in the EU. Generally, informal systems of coordination suffice to limit conflict and overlap. When these fail, for whatever reason, the Cabinet Office European Secretariat (or, to be more exact, one of the Cabinet's extensive network of committees) steps in to broker a common line across Whitehall. Normally, the coordination process in Britain works extremely well; so well, in fact, that the British representation in Brussels is widely regarded as 'a model of excellence' (Wallace, 1997, p.682; see also Metcalf, 1994). It is often said that these strong coordination mechanisms prevent departments from home running when intergovernmental discussions begin on a Commission proposal (Bender, 1991, p.20; Bulmer and Burch, 1998). The well established and respected culture of coordination and information sharing also provides each and every department with an adequate opportunity to flag potential implementation problems (i.e., 'misfits') when British negotiators from a cognate department are deciding whether or not to upload or download a particular EU policy proposal. Together, these informal and formal mechanisms have given Britain a reputation for 'negotiating hard' in the EU and then 'implementing well' (Wallace, 1995, p.47). If they are functioning properly, there should be very few misfits, and virtually no unforeseen misfits.

Like other Whitehall departments, the DoE recognizes and respects the importance of maintaining good internal coordination. An internal report prepared in 1993 concluded that coordination in Britain was 'better than most [other EU states]' and 'working well' (DoE, 1993c, pp.12–13). However, a 'Rolls-Royce' system of internal consultation and implementation does not necessarily translate into a positive and engaged relationship with the EU. The main reason is that, as discussed above, the political steer from the core executive has tended to be negative, as many Ministers (and senior civil servants) have, like the wider populace, been personally sceptical of the European integration (Menon and Wright, 1998). Similarly, very few British politicians have consistently and energetically championed environmental protection. These two factors certainly influenced the DoE's coordination of EU policy. An internal report (DoE, 1993c, pp.13–14) conceded that while EPEUR had effectively provided a good source of EU expertise and a means of coordinating EU policy, it was not so good at making proactive interventions (e.g., identifying 'uploadable' policy ideas) to ensure that 'wider strategic objectives managed by EPEUR, have a stronger influence on individual positions developed by other policy divisions [in the DoE].'

Problems with state-centric accounts

Since it was first developed, LI has been subjected to much criticism (see Wincott, 1995; Peterson, 1997; 1999; Jordan, 2001b/c) and has been subtly refined as a result. Six lines of attack directly concern the role of departments in facilitating and/or retarding integration and Europeanization. First, LI assumes that states function as a 'single agent' (Moravcsik, 1994, p.5), a view that neglects their internal diversity: departments fight against departments, and coalitional politics are rife in some states (M. Smith, 1999). This is a credible enough assumption to make if (like Moravcsik) the observer is only interested in IGCs and other 'big' decisions; but it becomes more suspect the more the analytical focus shifts down to the more mundane, day-to-day process of developing and implementing specific policies (see below).

Second, a number of key terms and concepts underpinning LI are unclear (e.g., 'state', 'state executive', 'national leader'). It is clear that Moravcsik (1994, p.4) is mainly concerned with the head of state ('chief executive') or Minister in a particular issue area but, as discussed more fully below, this risks overlooking the interpenetration of national and European bureaucracies in the EU. Rather confusingly, 'societal groups' are said to comprise interest groups and political parties, but also civil servants and other cabinet ministers. In other words, having assumed the state is unified Moravcsik promptly disaggregates it! More recently (Moravcsik, 2000, p.286), he has accepted that 'the state as a whole' is unhelpfully broad and offered 'the executive or government' instead, although most scholars of British government would dismiss this as being just as vague.[1]

Third, in spite of their importance, the precise conditions under which 'slack cutting' arises are very poorly theorized. On the one hand, LI argues that state preferences are determined by domestic societal pressure, yet, when these constraints are 'loose', the executive can 'shirk' tasks and pursue 'her preferred policies' in Europe (Moravcsik, 1994, p.5). According to the definitions supplied above, this could conceivably include one Minister 'home running' against another, although only by undercutting one of the central tenets of LI, the internal coherence of states. There is another important aspect to all this: if societal groups do not determine state preferences, where do they come from: from inside the state itself?

Fourth, the exclusive focus on 'big' decisions blinds LI to the more mundane policy shaping activities that take place in a multitude of institutional venues *after* (and before) the big bargains have been struck. These activities operate at what Joseph Weiler (Chapter 2) has described as a subterranean level in the EU. The media and most pressure groups do not routinely scrutinize decision-making at this level; even cognate departments struggle to keep up with what is decided. On this view, the big decisions 'represent the tip of an iceberg' of *informal* decisions (Armstrong and Bulmer, 1998, pp.56–7). According to Marks (1993, pp.392, 395), EU scholars need to go: 'beyond the areas that are transparently dominated by Member States . . . Beyond and beneath the highly visible politics of Member State bargaining lies a dimly lit process of institutional formation . . . [where] there is the less transparent but very consequential, process of post-Treaty interpretation and institution building.'

Fifth, even with the 'slack cutting' argument, LI struggles to explain why states – which (under an earlier elaboration of the model) are preoccupied primarily with 'safeguarding their countries against the future erosion of sovereignty' (Moravcsik, 1991, p.27) – are progressively undermining their own existence by diffusing state authority to other actors. Recently, Moravcsik (2000, p.286) has responded to this point by arguing that states willingly reduce their autonomy in order to achieve policy objectives. If that is so, surely the key questions are: (1) do 'states' (however defined) consistently and completely achieve their policy objectives (however derived) in the EU; and (2) if so, at what cost (however defined) to themselves? We return to these points in the case study chapters.

Finally, does any EU Member State approximate the rational, internally coordinated actor found in state-centric theories? In terms of inter-departmental coordination, yes, Britain is more coordinated and hierarchical than most EU states, but it is still deeply 'differentiated' (Rhodes, 1997). From the moment Britain joined the EU, a tension has existed between the centripetal forces of departmental pluralism (as discussed in the previous chapter) and the centrifugal forces exerted by the core executive. Formally speaking, UKREP is still an FCO-led mission, but nowadays it behaves much more like a quasi-independent 'mini-Whitehall' in Brussels, and leads the negotiation of British EU policy. The fact that most British departments are now 'thinking European' and have their own sophisticated and direct links with UKREP makes it more and more difficult for the core executive to police their activities in Brussels. Then there are also intra-departmental coordination problems. Even within a single department, conflicts routinely arise between units handling different issues (e.g., Europe/non-Europe, environment/'non' environment). The segmentation of policy-making even within one department makes it difficult for the European coordinating unit to assert control over the line divisions that upload and download items of EU policy. The DoE almost certainly experienced these problems. In 1993, an internal report remarked that: 'EPEUR has not been adequately resourced to offer an extensive service of EC consultancy across DoE, and . . . it is not a "natural reflex" for officials to seek advice from colleagues working in a different policy Group.' It continued: '[EPEUR's] position as a source of advice and expertise . . . is not consistently recognised and used by DoE officials working on [EU] business which is not concerned with environmental protection' (DoE, 1993c, pp.14, 13).

Process-based theories

March and Olsen (1999) describe LI as a *consequential* theory of the EU because it assumes that actors choose rationally between a number of courses of action whose likely consequences are well known, to secure the one that confers the largest net benefit. By contrast, process-based theories are more interested in the way integration and Europeanization unfold as a sequence of much smaller and, of themselves, insignificant decisions. When viewed over a longer historical timeframe, each decision to integrate can be seen to have engendered political support, thereby giving fresh impetus to the next round of integration. Consequently, political systems may 'drift great distances through cascades of modest steps' (March, 1981). Think of a lake, for example, gradually filling up with sediment. As the level of sediment rises, the fish are still able to swim but they have less and less water in which to do so.

Neofunctionalism

Historical institutionalism (HI) and neofunctionalism are the two most well known process-based theories of the EU. Neofunctionalism views integration as a cumulative and expansive process of integration, with a more or less automatic transfer of authority from Member States to EU institutions. Integration will, it is suggested, begin at the national level as states coordinate their functionally interconnected economic sectors (e.g., coal and steel). But further integration soon follows in its wake as elites shift their activities to the supranational level, followed by national pressure groups. These initial integrative thrusts are gradually cemented into place by the willingness of previously 'national' actors to transfer their activities, loyalties and expectations to the EU. (Neo)functionalism stresses the importance of *engrenage* (literally 'meshing in'), which occurs when actors from different national bureaucratic settings interact and, in so doing, readjust their preferences by 'learning' how to work together in new, supranational institutional contexts. Today, it is still common to find residues of neofunctionalist theorizing (e.g., Weale *et al.*, 2000), but it is rarely applied as a single, all-encompassing theory. Among other things, neofunctionalists greatly underestimated actors' allegiance to their national agencies and the 'tightness' of national coordinating mechanisms (Cram, 1997).

Historical institutionalism

Many aspects of neofunctionalism have been adopted by historical institutional theories of the EU. Historical institutionalism (HI) is *historical* in the sense that it tries to understand policy processes as they gradually unfold over time. History 'matters' because the constant reproduction of events is encoded in EU institutions, which in turn shape subsequent patterns of behaviour by altering the behaviour of actors. HI is *institutional* in the sense that institutions such as the Commission and the Court, as well as secondary rules (e.g., the *acquis*) are assumed to have an independent causal role in society. They codify the flow of history and constrain the opportunities for future change by altering the way preferences are articulated. In contrast to LI, both neofunctionalism and HI assume that state preferences are: fluid (i.e., they evolve in the period between 'big' decisions); sector specific (i.e., they depend on the institutional context of action); partly endogenously derived (i.e., they are affected by the steady expansion of the *acquis* and by EU bodies such as the Commission).

Process matters

HI contains a number of key concepts, which are widely agreed upon. First, it suggests that policy-making is better understood as a cumulative *process* rather than a series of one-off events. What happens in the past shapes future possibilities (Thelen and Steinmo, 1992, p.8). So, instead of rationally pursuing a set of strategic objectives, actors will often spend more time working under the 'dead weight' of previous institutional choices (Pierson, 2000b, p.493): 'history creates context, which shapes choice' (Aspinwall and Schneider, 2000, p.16). If we think about the 'big' IGCs, states are seldom free to select *de novo* from a long list of preferred institutional forms, since their choices are shaped and their options limited by the items of the *acquis* that have gradually accumulated in the period since the last 'big' decision (Cram, 1997). On this view, the 'grand bargains' simply codify what has already taken place (i.e., most integration occurs during the *intervening* periods). In other words, 'states' – to borrow from Marx – make intergovernmental decisions at IGCs, but in circumstances not entirely of their choosing. Similarly, once adopted, items of secondary policy can easily become institutionally entrenched and hence difficult to change. Because rules cannot be amended unless at least a qualified majority and (depending on the decision rule) possibly even *all* the states are in agreement, the status quo (however sub-optimal) will pertain. Pollack (1996, p.440) refers to this phenomenon as a 'joint decision trap'. States, of course, realize this, so they employ various devices to subvert Directives at the implementation stage (e.g., *de minimis* implementation) in order quietly to reduce 'misfits', rather than directly challenge the Commission or the legitimacy of the rules themselves. In so doing, the process of incremental policy accumulation substantially reshapes the very identities, self-images and preferences of the actors involved, as they realize what is or is not institutionally possible or appropriate. James March (1981, p.570) writes: 'actions taken by an organisation . . . become the source of a new definition of objectives . . . actions affect the preferences in the name of which they are taken; and the discovery of new intentions is a common consequence of intentional behaviour'. Chapters 7–10 explore the daily evolution of EU environmental governance and its impact on the strategies and interests pursued by the DoE, in order to test the claim that 'process matters'.

History matters: policy feedback, societal 'lock in' and path dependence

The past is also important because, once institutions are in place, actors adapt which effectively 'locks' them in place from below (Pollack, 1996, p.442). In the previous chapter we saw how British pressure groups and the Commission learned how to stop national environmental departments from subverting Directives by working together to flag instances of non-compliance. Societal lock-in represents another example of how 'policies produce politics' (policy feedback: see Pierson, 1993, p.597). That is to say, policies create new political opportunities, which in turn empower pressure groups. Societal adaptations (e.g., the construction of waste treatment facilities; the development of extensive legal systems) also make sudden policy changes unattractive. Consequently, policies will tend to be 'sticky' in the sense that they persist long after they cease to be an optimal response to the problems they were originally designed to address. What implications does this have for the behaviour of departments/states in the EU? It partly explains why 'states' engage in a regulatory competition to minimize misfits by uploading policies to the EU that they have previously developed at the national level. For instance, in the previous chapter we saw how the DoE fought to retain types and styles of national policy that, as an organisation, it felt comfortable with, while actively opposing the EU's attempts to download alien approaches.

Policy feedback can, of course, also constrain behaviour as well as enable it, as societal adaptations to past policies will, in time, feed back to the state, 'chang[ing] the administrative possibilities for official initiatives in the future, and affect[ing] later prospects for policy implementation' (Skocpol, 1992, p.58). If we think back to the early development of the environmental *acquis*, Britain fought to preserve its own, deeply-rooted system of pollution control. However, the DoE soon discovered that its informal, case by case approach to problem-solving was too unspecific with regard to burden-sharing and timetables of compliance to serve as a general model for the rest of the EU (Haigh, 1984, p.305), which requires clear deadlines and precise goals to achieve comparability of effort. With all the political will in the world, many aspects of British environmental policy were simply not uploadable to the EU.

Finally, policy feedback also affected the DoE's behaviour in one other significant respect. British policy had always been devolved to technical bodies, acting locally. So, when the EU began to discuss highly technical standards governing things such as bathing water and car emissions, responsibility naturally passed down to the line divisions and their associated agencies (e.g., the IAPI). The EPG could, of course, have created a European coordinating unit staffed with high-flying civil servants, who could have seized the reins of EU policy by uploading policies to Brussels, but it did not (see Chapter 2).

Policy feedbacks and lock-ins make institutions 'sticky' (i.e., insensitive to changes in actor preferences). In fact, the incentive to change may well decline as the returns reaped by remaining on a particular policy path increase over time: it is often simply easier to go forward than back. The presence of path dependence (i.e., the notion that policy is a development of what has gone before) has three important implications for the aims of this book. First, formative moments are critical. Decisions taken at the beginning may seem unimportant but they may decisively affect the course of history. Therefore, early successes and 'early accidents may be self-reinforcing' (Pierson, 2000b, p. 485). If we think back to Chapter 2, the European Council took a 'critical' decision at the 1972 Paris Summit to invite the Commission to develop ideas for EU environmental policy. Until then, the environment had been a matter of *national* policy. The DoE's failure to create a high-powered, strategically focused European co-ordinating unit was another early 'non' decision. At the time, these were perceived to be relatively 'small' decisions but, as we shall see, both have had enduring consequences (Pierson, 2000a, p.263).

Second, as EU rules develop in a policy sector, the more likely they are to clash (i.e., 'misfit') with pre-existing national institutional forms. In process-based terms, integration and Europeanization are, in effect, the outcome of a constant struggle between two institutionally rooted processes, one at the European level, and one at the national level. In Chapter 1 we noted that Europeanization arises when the institutional logic of EU policy triumphs and states are forced to adjust their pre-existing approaches. By contrast, states domesticate the EU when they successfully upload their preferred policies or when they subvert policies downloaded from the EU.

Finally, what circumstances are likely to produce sudden and dramatic institutional shifts (i.e., significant Europeanization)? HI argues that an accumulation of problems culminating in a political crisis is normally required to shift actors and organizations on to new paths (Peters, 1999, p.68). For example, Chapter 2 revealed how politicians such as Thatcher and Patten responded to the adverse publicity generated by the growing misfit between EU and national policies by ordering radical shifts in national policy. The empirical chapters show how the EPG benefited from the ensuing crisis, allowing it to upload more policy to the EU and adapt 'misfitted' national policies in areas such as water and air pollution control, to fit EU requirements.

The institutional context of action matters

In contrast to consequential theories, process-based theories argue that actors will alter their behaviour, identities and preferences to reflect what is institutionally appropriate in a given situation (Thelen and

Steinmo, 1992, p.8). If this were true, we should find states behaving differently when 'big' decisions are taken in IGCs from when 'smaller', more mundane decisions are made in the Environment Council (March and Olsen, 1999, p.952). In the big set-piece events, actors may perceive more to be at stake. Confrontation and posturing are permitted and indeed encouraged by the intense media interest. But in the Environment Council, Ministers normally take the lead and the norms of appropriate behaviour are usually more supportive of collective problem-solving (*engrenage?*). Over time, the participants (i.e., departments not 'states') *do* become gradually enculturated with a set of norms and expectations, which trigger a 'self-promoting dynamic of consent' (Héritier, 1999, p.58). In their review of EU environmental policy, Rehbinder and Stewart (1985, pp.265–7) detected a 'common value system that allows for a fair amount of harmonisation *even in the presence of conflicting economic interests*' (emphasis added).

In scholarly terms '[a]ctors enter into . . . relationships for instrumental reasons, but develop identities and rules as a result of their experience, thus shifting increasingly towards rule-based action, which they then pass onto subsequent actors' (March and Olsen, 1999, p.953). A participant in many negotiating sessions, Derek Osborn (1997, p.15), appears to confirm this when he describes the Environment Council as having a 'function-led dynamic' which is 'strongly pro-environment, and several Member States are thereby encouraged to make better progress than they would do by themselves'. In short, the institutional context of action does *matter*. Think back to the history of EU environmental policy described in Chapter 2: throughout the 1970s and 1980s European integration 'by stealth' generated a sizeable environmental *acquis* in spite of the prevailing economic and social conditions.

When we move down to the most humdrum issues, such as the negotiation of a Directive, we find middle-ranking civil servants working with very little direct input from elected politicians or cognate departments. It is here that we find the policy networks of special interests and interested specialists, which allow essentially 'national' actors to meet and play positive sum games (Peterson, 1995, p.391). Networks provide an especially fertile context for European integration (i.e., post-Treaty spillover). Moreover, the vertical divisions between different policy areas favour specialists (e.g., functional departments) over generalists (e.g., core executives: see Peterson, 1997, p.5). They facilitate home running and make central coordination harder to achieve. Hayes-Renshaw and Wallace (1996, p.15) probably exaggerate the influence of bureaucrats when they suggest that '70% of all EU decisions are taken in technocratic working groups', but Peterson's (1997) central insight – that the politicians who make the 'big bang' decisions do not participate directly and continuously in the day-to-day process of EU governance – is an important one.[2]

Institutional venues and norms of appropriateness

For advocates of a process-based view of the EU, the possibility that the institutional context of action might conceivably shape actor interests and strategies raises many intriguing questions about how individual departments go about uploading and downloading policy. Of course, this is not a new scholarly idea: '[n]ational interests will be defined differently on different issues, at different times, and by different governmental units' (Keohane and Nye, 1989, p.35). In other words, preferences are not exogenously but partly *endogenously* defined. 'Membership matters' (Sandholtz, 1996): domestic actors recalibrate their goals as a result of EU membership.

So, rather than treating states as though they were rational, calculating actors with anthropomorphic capacities, process-based theories suggest that we should instead study how their component parts (i.e. departments) operate in the various institutional venues of the EU (Peters, 1992; Marks, 1996, p.24). It is certainly conceivable that integration and Europeanization could advance the short-term departmental interests of individual departments of state. But, for advocates of state-centric approaches, the crucial question is whether these dynamics substantially weaken the control exerted by 'the state' in the longer term. Studies of the most deeply Europeanized policy sectors suggest that they do. Take Lindberg and Scheingold's (1970, p.6) study of post-Treaty dynamics in the agricultural sector:

> Ministers of Agriculture . . . have come to share preoccupations and expertise. They are subject to similar constituency demands, engaged in annual budget battles against their respective Ministers of Finance, and they seek the same general goals of improving the conditions of farmers and of modernizing agriculture. Indeed, in the eyes of many of their colleagues in other governmental ministries, they have come to form an exclusive club, thoroughly defended by impenetrable technical complexities. In a sense, the [Agriculture Council] . . . enjoys almost complete freedom from the inquisitive and restraining influences which each of the six individual Ministers has to submit to at home from his colleagues of Finance, Economic Affairs, and Foreign Affairs.

Clearly, departments do not simply go off to Brussels, negotiate what they like and then present the outcome to the rest of Whitehall as a *fait accompli*; that would be too blatant. It would also be very difficult in a country like Britain that has 'Rolls-Royce' coordinating mechanisms. In reality, European norms of appropriateness have to be reconciled with powerful Whitehall norms. The basic understanding among British officials that they should all 'sing from the same hymn sheet', even though the formal institutional machinery of coordination is quite light, is a particularly powerful 'logic of appropriateness' in Whitehall. Indeed, the DoE's own internal EU handbook argues that to be 'successful' (in departmental terms) at uploading and downloading, departments have to keep cognate departments 'on board' throughout the full cycle of policy development:

> The rationale for keeping other [departments] informed . . . is to try to avoid the introduction of new objectives to the UK position towards the end of a negotiation . . . [I]t may take considerable time and effort to have to do the thinking for other Departments but it is likely to pay dividends both in achieving your policy aims and avoiding major hassles in the end-game of the negotiations.
>
> (Humphreys, 1996, pp.37–8)

Nonetheless, if they are sufficiently determined and capable of 'thinking European' departments can 'achieve in the European arena what they [are] unable to achieve in the national context' (Toonen, 1992, p.111). Britain does have relatively strong coordination mechanisms, but they are quite reactive and rely heavily on individual departments showing self-restraint. In theory, this leaves room for what Keohane and Nye (1989, pp.41, 43, 44) famously termed 'transgovernmental relations' to develop. These arise when sub-units of government 'act relatively autonomously from higher authority in international politics' by building 'coalitions with like minded agencies from other governments against elements of their own administrative structures'. Golub (1997, p.14) finds no compelling evidence to suggest that that the DoE acted in this way, although he only sampled a small number of sub-areas of environmental policy prior to the SEA (i.e., before the most active period of EU policy-making). However, there is plenty of anecdotal evidence to suggest that departments do make 'home runs' (Barnett, 1982, p.133; Stack, 1983, p.149; Rehbinder and Stewart, 1985, p.316; Haigh, 1995a). Departments are able to 'exploit European links and alliances to their own advantage' (Burch and Holliday, 1996, p.90).

The case studies (Chapters 6–10) assess the extent to which uploading and downloading in the environmental sector was achieved by the DoE at the expense of other Whitehall departments.

Putting the pieces together

Pierson's (1996) theory of the EU tries to pull many of these elements of process-based thinking together. He agrees that state actors may be important when 'big' decisions are taken to create a policy area or EU organization. But before long, joint policies (institutions) begin to take on a life of their own, structuring and, on occasions, constraining the terrain upon which state actors (e.g., Ministers) attempt to amend existing policies and adopt new ones. Pierson is primarily interested in explaining integration, but his account is also directly concerned with Europeanization (although he does not actually employ that term). He offers four reasons to explain why integration and Europeanization may generate unexpected and undesired (at least by their creators) outcomes. First, state executives normally have very short time horizons: politicians are under electoral pressure to agree to policies with short-term pay-offs even when there is the risk of long-term costs and unintended consequences. In the crucial, formative moments, the core executive may be distracted, uninterested or preoccupied with short-term concerns. Chapter 2 revealed that many of the early environmental Directives were not actually taken that seriously by national environmental ministries (Jordan, 1999). Second, state preferences are not fixed: when looked at historically, states regularly change their preferences as a direct result of their continuous involvement in the process of EU policy-making (i.e., *process and context matter*: see above). Third, supranational policy-making is complex: state executives find it difficult to anticipate the long-term consequences of delegating authority to the EU. Finally, supranational actors are partially autonomous of states: they look for opportunities to extend their power and autonomy by exploiting gaps in state control and expanding 'misfits'. The Commission also extends misfits by employing 'subterfuge' (see Chapter 2) to achieve agreement at the policy-making stage. The case study chapters are replete with examples of the Commission making 'big' decisions look 'small' by breaking them into smaller, more palatable chunks, and using various other ploys to 'get its foot in the door' (Héritier, 1999, pp.58–9, 220).

Pierson then explains why states find it difficult to regain control of policies when their impacts (Europeanization) begin to unfold along an unfavourable path. First, supranational agents work hard to prevent states from reducing misfits by disavowing their commitments. In the following case studies, we will see how the Commission and the ECJ staunchly defend the environmental *acquis* in the teeth of state pressure. Second, decision rules (see above) produce a ratchet effect that binds states together, making it difficult to undo or revise the *acquis*. Finally, domestic societies adapt to the *acquis*, locking it into place (see above).

Problems with process-based accounts

Process-based theories have been critiqued on both empirical and theoretical grounds. Moravcsik (1998, pp.489–94), for example, staunchly defends his approach, believing that a detailed scrutiny of the biggest decisions shows that states *did* make conscious choices about the EU, which had foreseen and desired consequences. This book seeks to provide a more thorough test of that claim by looking not only at the policies adopted by the EU (i.e., integration) but the extent to which they produce the outcomes expected of them (i.e., Europeanization). It deliberately concentrates on a unitary, 'rational' but fairly sceptical state (Britain), examining how it managed a particular policy area (the environment) across a suite of big and small decisions.

It is undeniably true that process-based approaches do not (yet) add up to a coherent research programme (Rosamond, 2000, p.114), but much of Moravcsik's criticism of it to date has been made from within a state-centric paradigm (e.g., he argues that process-based approaches are deficient because they do not account for or predict 'state' preferences). However, more substantive theoretical criticisms could be levelled at process-based approaches. First, they appear only to explain periods of stability (or at least slow change). Sudden deviations from this pattern are simply explained away by sudden, exogenous events. Second, they are not as elegantly or coherently set out as LI. Some claim that they are not falsifiable and have low predictive power (Peters, 1999, pp.75–6). Third, advocates openly admit that the causes of slow policy change need further refinement (Thelen, 1999); critics also warn of excessive determinism (i.e., 'things happened in the way that they did because they had to': see Peters, 1999, p.75).

Conclusion

The dynamic interaction between national departments of state and European-level activities presents scholars with an analytical quandary because the processes at work are neither exclusively 'national' nor 'international'. While there is still no single theory (or even suite of theories) that provides a comprehensive account of the dynamic interaction between integration and Europeanization, two popular theories of the EU present sharply contrasting predictions about the behaviour of 'states' and their constituent departments. Put very simply, state-centric accounts are preoccupied with *formal* integration, (i.e., the 'big bangs' of European history); they assume states behave as single, rational actors regardless of institutional venue; and they believe that integration and Europeanization are sanctioned by states to suit national economic interests or (in some circumstances) the political concerns of the core executive. The purpose of this book is to understand if, when, why and how departments *matter*. For most advocates of a state-centric position, the answer to that question would be 'not much', not least because the mundane policy processes that they oversee are not as important as the issues determined by the big decisions.

In sharp contrast, process-based accounts suggest integration is mostly *informal* and 'evolutionary' insofar as it occurs between and as a context to the 'big' decisions (Armstrong and Bulmer, 1998, pp.54–6). This view brings departments to the very forefront of analysis as they negotiate much of the informal *acquis*. Process-based theories emphasize how 'irrationality, confusion and mistaken judgements' (Wallace, 1999, p.158) generate path-dependent dynamics that, crucially, steadily transform the tactics and political interests of those involved, including departments. On this view, individual departments *can* make a difference, although not always in expected ways. The next seven chapters analyse the origins and outcomes of 'big' and 'small' decisions through these two theoretical lenses, to determine which possesses the most explanatory power.

4
The Negotiation of the Single European Act

Today, the Single European Act (SEA) is widely recognized as having facilitated much deeper European integration across a suite of hitherto politically moribund policy sectors. In the environmental sector, it provided the *acquis* that had accumulated slowly and informally since 1970 with a formal legal underpinning (see Chapter 2). This was hugely important because, until the signing of the Act, the legal justification of EU environmental policy was always open to challenge (several influential commentators in Britain even questioned the *vires* of EU policy). The SEA removed these doubts and uncertainties by firmly embedding environmental issues in national and European level politics; it emphasized that EU environmental policy was finally 'here to stay'. This chapter considers the DoE's part in bringing about this important transformation in its capacity both as a negotiator of the Act (i.e., European integration) and a mediator ('fitter') of its long-term impact in Britain (i.e., Europeanization). It reveals that the Act transformed the political and economic fortunes of the EU across all policy sectors including the environment, to an extent that surprised even the most optimistic advocates of European integration.

The environmental provisions of the Single European Act

The Act introduced three important changes to the institutional rules governing EU environmental policy: (1) it provided an *explicit legal base* (namely Article 130r, s and t) in the Treaties; (2) it introduced *qualified majority voting* (QMV) for proposals linked to the completion of the internal market (Article 100a); and (3) it granted the EP the power to cooperate with the Environment Council in policy-making. As well as altering the formal decision rules of EU law-making, the SEA also had a hugely important symbolic effect on EU policy processes by establishing in EU law the ethical notion of environmental protection policy *for its own sake* rather than simply as an adjunct to the internal market. Crucially, the signing of the Act meant the Commission no longer had to justify each and every proposal in single market terms, allowing it increasingly to invoke a very different justification – that of achieving a consistently high level of environmental quality across Europe – for expanding EU competence. In this respect, it significantly shifted the terms of the debate away from '*should* the EU act?' in a given situation, to the narrower and much more technical question of '*how* and at what level should it act?' This served to accelerate the development of the *acquis* and facilitate the adoption of more stringent environmental rules (see Chapter 2). The fact that measures adopted under Article 100a (4) had to take as their base 'a high level of environmental protection' indicates how far the EU had moved on from its core agenda, which was to facilitate free trade by harmonizing product standards. Finally, Article 130r (2) of the new Act helped to embed environmental protection in the EU by requiring environmental protection to be integrated into *all* areas of EU policy-making via the principle of EPI (Jordan and Lenschow, 2000). At a stroke, this made environmental protection a matter for internal dispute within the Commission and the Council, and so raised political awareness of environmental issues to a much higher level among previously 'non' environmental departments and agencies in Brussels and national capitals.

Much has been written about the SEA, but there is still confusion about the provenance of its environmental provisions. One highly respected observer has claimed that 'the agitation of environmental bodies and others was transmitted via the European Parliament and resulted in a new treaty title on the environment' (Haigh, 1998, p.68). Moravcsik (1998), on the other hand, denies the importance of supranational bodies, arguing that the environmental provisions were just a side payment in a much bigger intergovernmental bargain between the three largest states – Germany, France and Britain – over the single market. The fact that these three were so preoccupied with negotiating these bigger, primarily economic, issues, allowed the Commission to 'quietly slip' in new Treaty powers relating to the environment and technology (Moravcsik, 1991, p.46). This does not, Moravcsik argues, mean that the Commission enjoyed unbridled autonomy to include whatever it wanted in the SEA. Far from it; the big three decided of their own free will to include environmental measures in the Treaty, having made a careful assessment of the likely implications. In short, they accepted the

Commission's plea to codify the environmental *acquis* because they believed it posed no serious threat to their autonomy. But which is the right interpretation: Haigh's more process-based explanation or Moravcsik's state-centric account? This chapter seeks to answer this question by examining the events leading up to and immediately beyond the signing of the Act.

The Single European Act: the 'biggest bang' in European Union environmental policy?

In Chapter 3 we noted Moravcsik's (1998) claim that integration normally proceeds as a sequence of irregular 'big bangs' rather than cumulatively and irreversibly through a series of feedbacks. Was the SEA a discrete 'big bang' or a small part of a much bigger process of change? Formally speaking, the SEA IGC ran for just over two months (early-September to mid-December 1985). The agenda was dominated by two main issues: (1) the Commission's plans for an internal market; and (2) changes to streamline decision-making (i.e., QMV) and facilitate or strengthen EU action in response to new problems (e.g., the environment) by adding new policy competences to the Treaty. In one of the most exhaustive analyses undertaken within the state-centric canon, Moravcsik (1991) argues forcefully that the outcome of the IGC, the SEA, represented the lowest common denominator of state preferences. In 'all other cases other than the internal market' the lack of consensus among the big three whittled the content down to 'a minor or symbolic level' (1991, p.42). Because of this, 'the British were most satisfied with the final outcome' he asserts (1991, pp.42, 49), although they were forced to accept (as part of a strategically bargained concession) the extension of QMV to internal market issues. Environmental issues without an internal market dimension would, he continues, remain under unanimity rule (1991, p.43). Throughout, he portrays the Commission as an insignificant player, although he credits it for having slipped in new Treaty powers relating to the environment and technology (1991, p.46). But this exception does not, he claims, undermine his overall argument, which is that states only accepted the need for a formal codification of the environmental *acquis* because environmental issues were already being addressed informally by the EU (Moravcsik, 1998, p.374).

Assessments made in the immediate aftermath of the IGC suggested that SEA would not amount to much. David Judge (1985, p.328) said it 'helped 'trip' the qualitative leap to European Union and . . . reduce[d] it to a stumble towards closely delimited cooperation'. Juliet Lodge (1986, p.221) described it as 'little more than a charter to safeguard and preserve national interests in the face of pressures for greater EC action in new and old areas'. And, in late 1986, the EP adopted a resolution condemning the final text as weak and wholly unambitious (Corbett, 1987, p.242). Members of the European Parliament (MEPs) were aggrieved that national leaders had not implemented the ambitious vision contained in the 'Draft Treaty on European Union' that the EP had produced in 1984.

Golub's (1996) state-based account of the environmental provisions of the SEA confirms many of Moravcsik's claims but qualifies a number of others. He emphatically denies that the environmental parts were quietly 'squeezed in' by the Commission without national politicians or environmental departments noticing. On the contrary, British negotiators accepted codification because they believed it not only accounted for, but actually enshrined, Britain's environmental policy style in European law (Golub, 1996, p.714). They believed they had closed the door to further Europeanization by successfully uploading British principles in the SEA such as subsidiarity, 'sound science' and cost-effectiveness (Wurzel, 1993, pp.182–5). The British core executive also expected the Commission to base future proposals on the environmental Article (130), where unanimous voting would apply, rather than Article 100 (the internal market article) where QMV applied. Finally, he finds no clear evidence that the DoE used the IGC to 'home run' against cognate domestic departments.

The empirical account presented below confirms the broad gist of Golub's argument. During the IGC, British negotiators did indeed try to place conditions on the use of Article 130r to render it a quiet *cul-de-sac* where some of the Commission's more grandiose proposals could be safely parked. However, Golub fails to provide a convincing riposte to the charge made by process-based theorists that '[i]t is almost always possible, *ex post*, to posit some set of Member State preferences that reconciles observed outcomes with the image of near total member-state control' (Pierson, 1996, p.125). In other words, what were Britain's policy preferences, where did they come from and to what extent were they satisfied? Golub is unclear on these points. First, he claims that state-centric theories explain the outcome of the IGC, only then to concede that the SEA had longer-term consequences (i.e., Europeanization) which 'somewhat disproved' Britain's expectations (Golub, 1996, pp.722–3). The remainder of this chapter traces the history of the IGC over a considerably longer timescale, and compares the original objectives of the various *parts* of the British government with the ensuing Europeanization of British policy.

The Inter-governmental Conference

Antecedents

The reasons why the then 12 Member States of the EEC decided fundamentally to overhaul the founding Treaties were many and varied. For the most part, they concerned economic and security issues such as the global competitiveness of European firms, the desire for economic reform and the willingness to address non-tariff barriers to trade (Moravcsik, 1998, p.293). Environmental concerns played little or no direct part in the decision to reform the Treaties,[1] although, paradoxically, the revisions did, as we shall see, have the positive (though not entirely intended) effect of transforming the rhythm and direction of EU environmental policy. In fact, the Act surprised even the most optimistic supporters of integration by transforming the political and economic fortunes of most EU policy sectors. But the SEA was not just a legal mechanism for achieving a single market; it was also a hugely complex inter-governmental bargain to improve the capacity and legitimacy of decision-making, increase the efficiency of European governance, achieve market liberalization and accommodate new members from the South of Europe. That the SEA was followed less than three years later by the opening of two more IGCs (Chapter 5) indicates the extent to which it transformed the political dynamics of European integration.

As one of the three largest Member States, Britain's support was absolutely vital to the realization of the single market. But a single market was *all* that Mrs Thatcher wanted. She most certainly did not want an open-ended IGC, which she suspected would lead to 'uncontrollable institutional ventures' (H. Young, 1998, p.330). However, at the June 1985 Milan European Council she allowed herself to be 'bulldozed' (Thatcher, 1993, p.551) into accepting just such a conference by a majority of European political leaders who saw deeper political integration as a necessary antidote to internal market reform. Indeed, elsewhere in her autobiography, she admits that: 'The pressure from most other Community countries, from the European Commission, from the European Assembly, from influential figures in the media for closer cooperation and integration *was so strong as to be almost irresistible*' (1993, pp.547–8, emphasis added).

A year earlier (June 1984) at a European Council meeting in Fontainebleau, she had won a long and extremely bitter battle with the EU over Britain's contribution to the common budget. After that victory her deeply held suspicions of the EU, which coloured every single department's perceptions of Europe, tempered. Things were going well for her at home and suddenly she began to feel that she actually understood the EU. Before, her 'automatic preference' was to resist integration by being confrontational (Gowland and Turner, 2000, p.261). However, the fatal mistake she made at Fontainebleau was to sanction the creation of a panel (the Dooge Committee) to discuss ways of reforming the EU to cope with enlargement and the internal market. It was, she recalls, 'one of those gestures which seem to be of minor significance at the time but adopt a far greater one in the light of events' (Thatcher, 1993, p.549). Among other things, the Dooge Committee eventually recommended an IGC to draft a Treaty of European Union, including new EU powers to protect the environment. The British representative, Malcolm Rifkind, entered a reservation on just about every single page of the Committee's report but by then the momentum towards an IGC was virtually unstoppable (Taylor, 1989, p.9).

The informal development of the environmental *acquis*

These were the 'high' politics that triggered the IGC. In order fully to appreciate how and why the environment entered the negotiations, process-based theories suggest that we go back and investigate the cumulative development of the *acquis* starting in the early 1970s. Recall that the *acquis* had developed informally since the 1972 Paris Summit (Chapter 2). Over the years, DG Environment had learned to make the most of the fairly limited legal foundations available to it, while recognizing the need to place the *acquis* on a sound legal (i.e., Treaty) basis. According to the then Director-General and former Head of CUEP, Tony Fairclough (1999):

> our goals were quite clear: to get something much stronger into the Act . . . and a part of the formal legal constitution of Europe where it could not be so easily challenged by those who didn't want an EU policy . . . [W]e would have liked even stronger requirements . . . but [the final outcome] wasn't bad.

DG Environment had, of course, always coveted a far bigger prize than the consolidation of the existing *acquis*, namely QMV. In its ten-year review of EU policy (1972–82), it claimed that unanimity, while not necessarily resulting in lowest common denominator outcomes, had failed to deliver sufficiently high environmental standards, and 'that in some cases stronger measures appeared to be justified'. It continued: 'the necessary search for unanimity is a slow process, taking up many man-hours of the limited resources available' (Commission of the European Communities, or CEC, 1984, p.16). So, as far as the Commission was concerned, a shift to QMV presaged higher environmental standards and a quicker and less resource-intensive route to achieving them. However, even as late as 1985, the likelihood of DG Environment achieving such a change was judged to be very low (Rehbinder and Stewart, 1985, p.334). The British core executive started out with much more minimal environmental expectations than the Commission (see Chapter 2). However, by the mid-1980s it too had come around to the view that the environmental *acquis* should be properly formalized. This subtle, but important, change of position was signalled in a key policy statement entitled *Europe: the Future* (HM Government, 1984), every single line of which had been personally scrutinized by Thatcher (George, 1994, p.176). Published just after the Fontainebleau summit, it set out her personal priorities, namely trade liberalization through international mechanisms such as the General Agreement on Trade and Tarriffs. But the document also accepted that the EU did have 'an important role to play' in the expansion of environmental protection (HM Government, 1984, p.75). A few months before, Thatcher had hosted a series of environmental policy briefings at Chequers aimed at improving the Government's European image at home and internationally (see Chapter 2). Thatcher desperately needed European allies to obtain a budget settlement. One way of achieving this was to court the Germans by improving Britain's environmental performance, particularly in relation to acid rain (ENDS, 112, p.3).

Preparations in Whitehall

Having being unceremoniously bounced into an IGC that she did not want, Thatcher could conceivably have stood aloof, but the Cabinet and her closest political advisers persuaded her to take part. They believed that British firms would thrive in a single market and were anxious not to be excluded. But first, Britain had to stay in the negotiating game, even if it meant accepting common policies in the so-called 'flanking' policy areas (such as technology and the environment) to achieve market liberalization. The Cabinet apparently had 'serious misgivings' (Taylor, 1989, p.8) about venturing too far into these areas, but on balance felt they would not seriously imperil Britain's economic and political interests (George, 1994, p.183). This was certainly the strong impression that Thatcher conveyed to British MPs immediately prior to the IGC (i.e., that it would not take long and might well fail completely: HC Debates, Vol. 82, 2 July 1985, cols 185–94). She calculated that it was better to participate than to exclude herself (Moravcsik, 1998, p.364). Her Foreign Secretary, Geoffrey Howe (1994, pp.407, 408–9), was convinced that: 'once an IGC had been convened, any and every possible "constitutional" amendment could be

proposed and discussed. Pandora's box would indeed have been opened
. . . But soon the irritations of Milan faded into history . . . we were in
truth eager to hammer into place the . . . single market programme.'

Britain did not (as is now the norm) publish a White Paper setting out
its objectives in advance of the IGC, but Howe did set out Britain's aims
in a statement to Parliament in July 1985. In that, he identified the internal market and closer collaboration on international affairs as major
priorities, but warned that concessions might have to be made on two
issues (cooperation with the EP and the creation of a 'European Union')
to satisfy other states (HC Debates, Vol. 83, 24 July 1985, cols 1061–70).
Up until this point, Britain's political position had been shaped almost
entirely by the core executive. Although line departments stood to be
affected by commitments made during the IGC, they were apparently
not consulted (Budden, 1994, p.265). Only later did the Cabinet Office
instruct the various line departments to assess the implications for their
own policy areas of shifting to different decision rules. It was 'rapidly
concluded that on virtually every sensitive issue where [QMV] was proposed it would not cause any problems . . . and that in some cases British
interests would be enhanced' (Wallace, 1996, p.64).

The IGC commences

The IGC itself has been described as a rolling 'pantechnicon': states
eagerly threw in their preferred ideas but very few fell off (McAllister,
1997, p.123). The British, however, were only very weakly engaged in the
early stages; FCO officials maintained a studied detachment, willingly
debating other states' proposals but withholding any positive suggestions of their own (Corbett, 1987, p.244). The environment was not,
however, quietly 'slipped in' to the space thus created. On the contrary,
the Commission employed the Monnet method and selected environment specifically because it was perceived to be uncontroversial and
hence immediately 'agreeable'. By securing agreement on the need for
codification, the Commission managed successfully to kick-start the
IGC by injecting a dose of much-needed confidence (Dinan, 1998,
p.29). These early interventions 'radically transformed the atmosphere'
of the IGC from one of open confrontation to constructive engagement
(R. Dehousse and Majone, 1989, p.102). However, the Commission's
masterstroke was to secure a foothold for environmental policy in the
internal market section of the Treaty (i.e., Article 100a: *Agence Europe*, 30
July 1985, Europe doc. 1366) governed by QMV. Thatcher was instinct-
ively suspicious at this, but considered QMV a price worth paying to
achieve the internal market. She apparently dismissed the threat of

future spillovers into flanking areas such as the environment as 'simply rhetorical' (Moravcsik, 1991, p.41). Moravcsik was not, however, the only commentator fundamentally to underestimate the long-term significance of the new environmental provisions in the new Act. The normally reliable ENDS Report concluded that the Treaty 'appears to have emerged, in line with the UK's wishes, little changed from the reform process' (ENDS, 131, p.22), with 'only the faintest glimmer of change' in the formal decision rules.

The endgame

The available documentary evidence – confirmed in interviews with senior civil servants – shows that the DoE was condemned to observe these unfolding events from the sidelines in London. MINIS 7 (July 1986) reported that '[h]eavy (and *unforeseeable*) demands' on the department had arisen during the IGC, underlining the reactive nature of its input (DoE, 1986c, CDEP, 2.03, p.9; emphasis added). Nigel Haigh, who had the ear of many senior civil servants in EPINT, believes that:

It was the first [IGC] and nobody really knew how they worked . . . [It] all happened by subterfuge. It was an intergovernmental matter . . . run by the Foreign Office . . . DoE officials did not know what was going on and did not have any role. The DoE certainly wasn't pushing for change . . . It was all a *fait accompli* . . . stitched up at a very high level without anyone in the DoE really knowing.

(Haigh, 1999)

Crucially, the then Head of the CUEP's EPINT, Fiona McConnell (1999), maintains that: 'We did think a new treaty was important . . . [which was also supported] . . . at a high political inter-departmental level, but only so long as it didn't affect the veto. Everything was done to make sure that majority voting didn't come into the environment [part of the Treaty].'

The environmental departments in some of the more environmentally progressive states, such as Denmark and The Netherlands, certainly supported some of the Commission's proposals (Wurzel, 1993, p.184),[2] yet the DoE implacably opposed QMV. Why was this? In Chapter 2 we noted the DoE's reluctance to embrace the European environmental agenda. Martin Holdgate explains that it was simply not willing or able to pool sovereignty at this time:

We disliked the idea of QMV altogether because we were still frightened of being railroaded by continentals with a different philosophy

... At the time we were fire fighting and QMV would mean more fires and more costs. As the department was a broad one my job was to keep the fires back down so the issues like planning and local government finance could receive the most attention. Our job was to damp down all the fuss about pollution killing you and poisoning the sea.

(Holdgate, 2000)

Crucially, Holdgate confirms that around this time, the EPG was much more likely to perceive the EU as a source of policy *problems* than a means of achieving higher environmental standards in Europe or in the UK.

Therefore, officials in EPINT worked with UKREP to draw up a list of rules to limit the scope of Article 130 to 'safe' environmental issues. However, Britain's efforts to delineate the precise limits of EU competence failed because other Member States were reluctant to deny themselves an opportunity to tackle new types of environmental problem in the future. Consequently, the IGC agreed to define the aims of EU environmental policy very broadly (de Ruyt, 1987, p.214; Vandermeersch, 1987, p.413). The Commission, which had also sought clear guidance, happily settled for a series of broad objectives, which left the precise limits of EU competence tantalizingly unclear (Krämer, 1987, p.664), but eminently exploitable at some future date.

Preliminary reactions

At first blush, the environmental amendments appeared to *solve* more domestic problems than they created. First, they promised to place a strong legal straitjacket on the future development of the environmental *acquis*. Long before the SEA was even thought about, their Lordships (HOLSCEC, 1978; 1979) had complained bitterly about the 'irreversible removal of legislative power' arising from nearly a decade of European integration, and implored the DoE to delineate the scope of future EU activity (see Chapter 2).

Second, there was very gentle pressure for codification from the EEB and the IEEP in Britain but, as we saw in Chapter 2, most British environmental pressure groups were simply not attuned to European developments. Consequently, the IGC passed them by. Their *de facto* 'European expert', Nigel Haigh, admits that the SEA 'was drafted quickly and with little discussion' (Wilkinson, 1990, p.i) with the DoE or its network of pressure groups. Third, Britain had become much greener in the period since 1973 and more receptive to tougher environmental standards. Europeanization was an important factor behind this step change in official and political attitudes (see Chapter 2), but so too was domestic political pressure, transmitted through the activities of environmental groups such as Greenpeace and FoE. By 1985, even Eurosceptical politicians such as Ridley had realized that Europeanization had to be engaged with even if it could not be resisted (Chapter 2). Finally, during the mid-1980s the *acquis* began to develop its own internal dynamism. For instance, during the preparatory phase of the IGC, the ECJ issued a landmark ruling on waste oils (Case 240/83), which described environmental protection as one of the EU's 'essential objectives' (Köppen, 1993, p.135). In so doing, the Court effectively blessed the IGC's 'big' decision to codify the *acquis*.

After the signing of the SEA, Thatcher reassured MPs that the 'amendments on environment and technology show more precisely the activities which were previously taking place' (HC Debates, Vol. 88, 5 December 1985, col. 436). The Foreign Minister who negotiated much of the text, Linda Chalker, even claimed that Britain would 'be able to ensure that, where we wish, decisions continue to be taken by unanimity' (HC Debates, Vol. 93, 5 March 1986, col. 338). Later, she strenuously denied that the SEA would facilitate deeper Europeanization:

> The Commission could put forward a proposal, but we are protected by the unanimity rule. The Commission is unlikely to do anything like that because we have worked in tandem with it as partners and it would not put forward something that other partners as well as us ourselves were not prepared to have. If we were not prepared to have it, we would not accept it and our partners could do likewise and it would fail.
>
> (HC Debates, 10 July 1986, col. 539)

Even the Environment Minister at the time, William Waldegrave, admitted to being 'a little gloomy as to whether in the immediate future the impact of the [Act] will actually be beneficial on the environmental side' (HOLSCEC, 1987, p.89).

After the Inter-governmental Conference: informal integration re-asserts itself

However, his civil servants evidently had a lot less faith in the legal firewall that the new Treaty had erected between internal market and environmental issues. In 1987, EPINT warned that 'a *major imponderable* is the effect on Environment Council business of the [SEA]' (DoE, 1987, Central Directorate for Environmental Protection (CDEP) 2.04, p.21; emphasis added). The following year, it again warned that the '[u]ncertainties over the use of the [SEA] . . . will continue' (DoE, 1988b, p.18).

Its concerns proved to be well founded; for instance, in the Fourth Environmental Action Programme (1987–1992), which was published within months of the completion of the IGC, DG Environment claimed that it had no option but to resort to Article 100a in order to avoid creating barriers to trade (CEC, 1987, para. 1.2.3).

In 1988 (i.e., a year after its entry into force), the SoSE, Nicholas Ridley, finally commissioned the IEEP (i.e., *not* his own civil servants) to examine the impact of the SEA on Britain and specifically Whitehall (Haigh and Baldock, 1989, p.4). The IEEP concluded that the SEA would not produce a 'single dramatic change', but warned that:

> To view Articles 130R, S and T as simply legislating the competence that the EC had de facto acquired in the field of the environment would be to undervalue them. They have symbolic value and their effects will be subtle. These subtle effects can best be felt and understood by recognising that for environmental purposes the EC is now a federal system.
>
> (Haigh and Baldock,1989, p.24)

The DoE was slow to recognize and respond to the informal integration that followed in the wake of the SEA (Chapter 2). In the years following ratification, more environmental legislation was adopted than in the previous 20 years, and the speed at which individual policies were agreed also increased (Jordan, Brouwer and Noble, 1999). Crucially, only half the new statutes were based partly or solely on Article 130 (Golub, 1996, p.716). The SEA also made it much easier for the EU to act in areas such as access to environmental information where the internal market dimension was either weak or non-existent. Moreover, the SEA gave the Commission the *moral* authority to intervene in these and other seemingly 'domestic' issues, such as urban wastewater treatment. In March 1990, the House of Commons Foreign Affairs Committee concluded that the Act had had 'much greater institutional impact than anyone predicted when it was ratified' (HC 82-I, 2nd report).

Geoffrey Howe admits to having being surprised by the Europeanization generated by the Act:

> The [SEA] was to achieve most of what we had hoped of it . . . [but] . . . we found ourselves facing on some social and environmental matters a more extensive use of Community powers than we had regarded as foreseeable or legitimate. Both these trends can probably be attributed to the impact of the [SEA] upon what might be called the

'culture' of the [EC] and its institutions. The habit of working together . . . which we had so strongly urged for the Single Market was not something that could be ruthlessly confined.

(Howe, 1994, pp.457–8)

Mrs Thatcher (1993, p.556) also concedes that 'the new powers the Commission received only seemed to whet its appetite'. Over the summer of 1988, she asked the Cabinet Office's European Secretariat to conduct a *post mortem*. It told her that the Commission:

sets up 'advisory committees' whose membership was neither appointed by, nor answerable to, member states and which tended therefore to reach *communautaire* decisions. It carefully built up a library of declaratory language . . . in order to justify subsequent proposals . . . But, most seriously of all, it consistently misemployed treaty articles requiring only a qualified majority to issue directives which it could not pass under articles which require unanimity . . . We did indeed fight and won a number of cases . . . before the . . . ECJ. But the advice from the lawyers was that in relation to questions of Community and Commission competence, the ECJ would favour 'dynamic and expansive' interpretations of the treaty over restrictive ones. The dice were loaded against us.

(Thatcher, 1993, p.743)

These are, of course, precisely the devices (or 'subterfuges': see Chapter 2) that process-based theories suggest are used by the Commission to extend informal integration stealthily and subtly. States tried, as Thatcher explains, to regain control of European integration by contesting the legal bases of some of the new environmental Directives that emerged after the SEA (R. Dehousse, 1998, pp.155–6). In a landmark case relating to pollution from the titanium dioxide industry, the ECJ sided with the Commission against the Council by deciding that the Directive in question should have been based on Article 100a rather than 130s (Case C-300/89). In effect, it supported the trend towards more QMV. According to Fiona McConnell (1999):

Ministers . . . were lulled into a false sense of security about 130r. Some of us [in the DoE] saw it coming but were assured diplomatically that 'no, no, no that wouldn't happen' . . . It was quite obvious to most of us – particularly the department's lawyers – that the politicians didn't want to be seen to be obstructing the single market drive . . . [The] trouble [is] when you're at the coalface, miners don't normally get up the mine to see the national head of mineworkers, never mind the minister of mines. I would say it suited the country to find areas where more QMV could be accepted without harming our interests . . . We [the DoE] were quite pleased to have the Treaty . . . but no not QMV.

In the course of time, the DoE responded by readjusting its *modi operandi* and philosophy (Chapter 2) by holding more meetings with MEPs, creating EPEUR and changing its negotiating style in Europe. However, these changes were only ever a reaction to EU policy-making, brought about when the *whole* department (i.e., *not* just EPG) finally realized that it was, to quote a former Permanent Secretary, 'being hit amidships' by a growing catalogue of misfits. Then and only then did the DoE realize fully that it 'had to raise its game in Europe' or suffer the consequences at home and in the EU. But as we saw in Chapter 2, before the DoE could engage more positively with the EU, it had to do what it had not done before; namely, to admit to the outside world and to itself that Britain had been Europeanized and that the EU could be steered, but not blocked indefinitely. This was the sub-text of a low-key speech made by Derek Osborn to a conference of public administrators in 1990, in which he observed that:

[i]n a relatively brief period of time the nature of the [EU] has evolved and expanded, so that its many strands are now woven into most areas of [UK] public policy. This is certainly true of the environmental policy, where in the last 20 years the [EC] has rapidly become a major factor, or even the dominant one.

(Osborn, 1992, p.199)

The SEA played an important part in facilitating the deeper integration that created the 'misfits' that continued the Europeanization of British policy. However, the most puzzling aspect of the whole saga is how little direct influence the DoE had on the wording of the environmental parts of the SEA.

Theoretical reflection

What has this chapter revealed about the ability of different actors to manage European integration and Europeanization? With hindsight, the British core executive does not emerge particularly well from the saga. Mrs Thatcher's preferred plan for an internal market had to be hurriedly re-crafted when she was embarrassingly outvoted at Milan.

Having lost the hard fought initiative it had secured with 'Europe: the Future', the core executive then found itself buffeted by external events that were neither predictable nor of its immediate choosing. Nonetheless, the immediate outcome was broadly in line with the predictions made by state-centric theories. That is, by acting as a single integrated unit in Europe, Britain *appeared* to have secured an internal market with only a number of seemingly minor concessions. In late 1985 the prospect of what Mrs Thatcher termed 'uncontrollable institutional ventures' seemed very distant.

Throughout the IGC, the 'state' appeared to function as a single, rational agent, with little evidence of the DoE 'home running'. On the contrary, the DoE was actively excluded from the 'statecraft' decisions taken by Thatcher and her inner cabinet. Throughout and immediately after the IGC, the continuing absence of focused societal pressure allowed the core executive to play 'two-level games'. For example, having signed the SEA the core executive deliberately understated its consequences, pushing it through the British Parliament under a guillotine procedure. Until the early part of 1986, the British core executive appeared to have controlled unwanted Europeanization by managing European integration, whereas the DoE was very much a cipher. Importantly, the DoE actively *opposed* changes that would have deepened integration and facilitated the Europeanization of domestic policy, but was overruled.

So, game, set and match to state-centric scholars? Not quite, because if we adopt a process-based perspective and fast-forward a decade to the mid-1990s, the picture looks very different. By then, the SEA was being implemented, the locus of policy development had returned to the functional level of the Environment Council and unforeseen consequences had begun to arise from specific items of secondary legislation. The speed at which Eurosceptics in the British Parliament and (even) the core executive moved to condemn the SEA is testimony to the powerful, integrative thrusts unleashed by the IGC (see, for example, Thatcher's Bruges speech in September 1988: Thatcher, 1993, pp.744–5). A process-based account correctly identifies the critical importance of the core executive's short time horizon, typified by Mrs Thatcher's unwitting commitment to the sort of 'unstoppable institutional venture' that she herself abhorred but from which she did not wish to be excluded. The paradox is that none of this was specifically sought or facilitated by the DoE. In fact, the prime mover was probably the Commission, which could afford to take a longer-term view of events than the DoE. There was a widely held expectation within many British departments (i.e., not just DoE) that Article 130 would be the main wellspring of future EU environmental policy. Legal experts from the Commission and the Member States had worked on this Article for the best part of six months, so there was, as one senior DG Environment official explains (Krämer, 1987, p.664), an understandable 'tendency to underestimate' the importance of Article 100a. However, the Commission, aided and abetted by the Court, soon learnt how to bypass the Article 130 *cul-de-sac* by subtly redefining its proposals as internal market measures.

The SEA was, of course, vitally important in this respect (it made the decision rules more *communautaire*), but so too was the wider political context of the late 1980s, which was considerably more supportive of higher environmental standards. Policy feedback effects helped to sustain integration as the political demand for higher standards fed back into the Environment Council, which began to produce environmental rules at a faster rate. Indeed, a process-tracing approach reveals that the greening of British politics in the 1980s was itself greatly intensified by the Europeanizing influence of earlier environmental Directives (see Chapters 2 and 7). In the late 1980s, European Directives began to create new political opportunity structures at the sub-national level in Britain and within Brussels (Chapter 2). These were eagerly exploited by domestic pressure groups to achieve higher domestic standards, thereby emphasizing the self-reinforcing character of European integration. Later on in the 1990s, British politicians from both ruling political parties (e.g., John Gummer and Michael Meacher) helped to lock the policy changes in place by gladly accepting the political credit for the higher levels of environmental protection that their predecessors had (sometimes unwittingly) introduced. Thus, it was the momentum generated by the greater use of QMV *coupled to* Member States' increasing appetite for environmental measures that helped to generate a strongly pro-environment dynamic in the Environment Council. All these aspects of the DoE's behaviour are consistent with a process-based view of the EU.

Moravcsik now accepts that supranational actors played a much greater role than in any other IGC (Moravcsik, 1998, p.347), though he steadfastly maintains that the 'main outlines [of the SEA] were firmly set by enduring national interests'. This chapter directly challenges his interpretation of the environmental provisions and reveals that a much more paradoxical combination of processes were at work. First, weak domestic pressure from environmental groups allowed the British core executive to 'cut' slack, but this had consequences that contradicted its initial purpose, namely achieving internal market reform (European integration) *without* accelerating the Europeanization of national policy in the so-called flanking areas. Second, Moravcsik (1998, p.365) also overlooks the events that occurred between the SEA and the next 'big bang'. Thus, in his chapter on the SEA, he argues that environmental policy 'remained under unanimity'. But 90 pages later (p.455), he writes that the Maastricht Treaty achieved little more than a 'reclassification of activities already conducted by QMV under Article 100'! Thus, by looking solely at the 'big bangs', he overlooks the intervening process through which environmental policy was gradually but substantially transformed by the daily decisions taken in sectoral policy networks. Finally, a process-centred approach demonstrates that the environmental *acquis* developed a life of its own in the mid- to late 1980s (i.e., well before the SEA IGC). Moreover, the preferences that the British core executive and the DoE took with them into the SEA IGC were not exogenous, as LI implies: they were shaped *endogenously* by the informal development of the *acquis*.

In taking a brief and unrepresentative 'snapshot' of a gradually unfolding process, state-centric theories seriously underestimate the potential for integration to move *beyond* the original (in our case *c.*1984) preferences of states and impose limits on their ability to fight back (e.g., the Environment Council mounted legal challenges and the ECJ resisted them). In the lexicon of more process-based theories, Britain gradually found itself locked into an expansive process of integration and Europeanization that was not of its immediate choosing. A process-based perspective reveals other important ways in which European institutions perpetuated themselves. Thus, state-based theories argue that Member States are separate from and in control of the process of integration; yet, in the creation of EPEUR, we have evidence of the British state being subtly changed by the very process it is meant to control.

To summarize, this chapter confirms LI's reputation as a sophisticated and parsimonious theory of intergovernmental bargaining during the set-piece political events, but it raises further doubts about its ability to account for the dynamic interaction between integration and Europeanization over an extended time period. Far from operating on a *tabula rasa*, negotiators faced an institutionally constrained choice between doing nothing and blessing the integration that had already taken place in the environmental sector. This, *contra* state-centric theories, probably lulled the core executive into a false sense of security. Second, the dividing line between day-to-day decision-making and the SEA IGC was far from clear-cut. Studying IGCs reveals the bargaining between core executives, but it does not fully explore the origins of state preferences or the subsequent process of interpreting and applying new Treaty mandates in functional policy sectors where the institutional dynamics and logics of 'appropriate' behaviour are somewhat different.

5
The Negotiation of the Maastricht Treaty

Thatcher signed the Single European Act believing it would promote 'Thatcherism on a European scale'. But as soon as she thought she had mastered Europe, the rest of the EU set off on a new track towards deeper European integration. Margaret Thatcher had always suspected that the SEA would eventually trigger further 'uncontrollable institutional ventures' (Chapter 4), but she signed the Treaty all the same. And so it did because, three years after its entry into force, the EU opened two more IGCs on political (IGC-PU) and monetary union (IGC-MU) respectively, to prepare for deeper European union. Why, then, did the EU embark upon yet another round of Treaty changes so soon after the ratification of the SEA? The simple answer is that the SEA contained a formal requirement to hold *an* IGC by 1992 to review common foreign policy-making. However, unforeseen political events, namely the fall of the Berlin Wall and growing support for monetary union, brought the start of the two IGCs forward by several years. These events raised new political problems, which were fed quickly and chaotically into the two IGCs. In no time at all, the political agenda of the discussions became hopelessly overloaded as different actors used the opportunity to push their pet projects and policy problems.

This chapter is primarily concerned with exploring the extent to which the DoE shaped the environmental agenda of the IGCs and the Europeanization that flowed from them. It suggests that like the SEA, the Treaty on European Union (TEU) was primarily motivated not by environmental, but *macro-political* considerations, although pro-environment actors quickly exploited the IGC further to embed the environmental *acquis*. We shall see how and why the Maastricht Treaty built upon and greatly extended the environmental provisions of the SEA in several important respects. Unless otherwise indicated, the remainder of this chapter focuses on the IGC-PU where the discussions about environmental rules took place.

Political antecedents

The primary stimulus for the TEU was internal, namely the desire to adopt a common European currency. However, the fall of the Berlin Wall in 1989 and the emergence of the 'German question' injected fresh urgency to the discussions about Europe's future. The prime movers were France and Germany: France because it wished to tie Germany down politically, and Germany because it wanted to embed economic union in a politically and constitutionally united Europe. The first steps towards establishing an IGC were taken at the June 1988 Hanover European Council, when Commission President Jacques Delors was asked to chair a committee to assess the necessary steps to European Monetary Union. However, states with more strongly pro-integration agendas soon pushed plans for a much more wide-ranging review of EU powers, which of course Mrs Thatcher vehemently opposed (George, 1994, p.195). As before, the rest of the British core executive eventually managed to persuade her to adopt a more conciliatory line (1994, p.217), but the fall of the Berlin Wall and Helmut Kohl's seemingly unstoppable quest for rapid unification served only to reaffirm her inability to steer European political affairs.

France and Germany hurriedly put together a joint plan to proceed with not just one but *two* parallel – though separate – IGCs (Baun, 1995, p.613). Mrs Thatcher managed to buy some time, believing that the EU should concentrate on completing the internal market (Brewin and McAllister, 1991, p.413). But finding herself more and more isolated, she eventually conceded defeat at the April 1990 European Council in Dublin (Dublin I: George, 1994, p.220), where she was outvoted eleven to one by the other European leaders (*EC Bulletin*, 4, 1990, point 1.12). So as not to offend her political sensitivities, their decision was not formally adopted until the Dublin (II) European Council in June. There then followed a gap of about six months before the formal opening of the IGCs in December 1990. Three important events occurred in the intervening period which completely transformed the negotiating environment. The first was the reunification of Germany (October 1990), which increased the demands for deeper political union. The second was Iraq's invasion of Kuwait in August 1990, followed by Operation Desert Storm. This proved to be a huge distraction for all concerned. Finally, there was the sudden resignation of Thatcher and the arrival of her more emollient and (initially) Europhile successor, John Major, just over two weeks before the commencement of the IGC negotiations. On coming to power, Major tried to heal the deep rifts within the Conservative Party which had precipitated Thatcher's downfall, by developing a more constructive relationship with Britain's European partner. However, his failure to unite the party and the Cabinet tied his hands at crucial points during the IGCs. The 'straitjacket of party controversy' (Wallace, 1995, p.52), though not nearly as constraining as it would later become during the ratification stages (see Baker, Gamble and Ludlam, 1993), did play a critical part in shaping Whitehall's handling of the IGCs.

The length of the IGC-PU – from December 1990 to December 1991 – reflected the complexity and fluidity of the agenda under discussion. The SEA IGC lasted just two months and fell neatly within one Presidency, whereas the Maastricht IGCs lasted a full 12 months, spanned three separate Presidencies, and were agreed in December 1991 but were not fully ratified until 1993. Even though Thatcher's departure increased the prospects of reaching agreement, consensus was often elusive and the pressure to agree *something* in time for the endgame at the Maastricht European Council in December 1991 remained extremely acute. Finally, in contrast to the SEA (where Britain helped shape the internal market agenda), the Germans and the French almost completely dominated the preparations for and shaping of the Maastricht IGCs. Wracked by Cabinet splits and intense infighting, Major's divided government was left struggling to extract whatever last minute concessions it could from the endgame.

The environmental provisions of the Maastricht Treaty

As with the SEA, environmental issues were an incidental aspect of the IGC. Nonetheless, the TEU did make a number of important changes to EU environmental policy. These included the following elements:

1 Amending Article 2 to make the promotion of 'sustainable and non-inflationary growth respecting the environment' one of the EU's core tasks. This replaced the Treaty of Rome's commitment to achieve 'a continuous and balanced expansion' of economic activities.
2 Extending QMV to most, though, crucially, not all areas of environmental policy, and introducing co-decision-making with the EP for measures linked directly to the internal market.
3 Strengthening the principle of EPI (see Chapter 4) in Article 130r(2) by *requiring* the EU to integrate environmental protection requirements into 'non' environmental policies.

4 Introducing the precautionary principle in Article 130r(2), thereby allowing the EU to address environmental problems before a firm causal link had been established by scientific evidence (see Wilkinson, 1992).

Advocates of a state-centric view of the EU have dismissed the IGC-PU as a 'sideshow' to the IGC-MU (Moravcsik, 1998, p.447), as being merely a 'concession' to the German government that needed to legitimize the single currency for an apprehensive domestic political audience (Baun, 1995, p.619). Worried that it could be excluded from a central 'core' of pro-integration states, Britain decided to engage fully in the IGCs although it continued to oppose deeper integration across the board (Moravcsik, 1998, pp.386–7, 462). In contrast to the SEA, state-centric accounts claim that supranational entrepreneurship was weak and sometimes completely counterproductive; the Presidencies and the Council Secretariat were much more effective at shaping the final outcome of the negotiations than the Commission (Laursen, 1992, p.244). Finally, because the majority of Member States were less keen on political union than Germany or disagreed about the form it should take, the outcomes of the IGC-PU were less ambitious and focused than those of the IGC-MU.

On the other hand, process-based accounts characterize the new aspects of the TEU as a political spillover from the commitment made in the SEA to achieve a single market (Sandholtz, 1993). The basic claim here is that the informal integration triggered by the SEA altered the preferences of states, which were then cajoled into deeper integration by supranational actors such as the Commission. In other words, process and history 'mattered'. States did not enter the IGC with coherent or fully-fledged preferences, but developed and redeveloped them throughout the process of the negotiations. Furthermore, even in this, the biggest of 'big' decision-making situations, the British state did not function as a single, rational actor. More often than not, party political (namely, the need to unify the Cabinet) considerations dictated the core executive's handling of the IGCs. 'Britain' was an unstable coalition of individual politicians (a 'they' rather than an 'it': Forster, 1998, p.358).

Prevailing accounts of the environmental changes made by the TEU adopt one of these broad perspectives. Moravcsik (1998, p.55), for instance, dismisses them as one of a number of 'minor' issues that 'went the Commission's way'. Like the changes made in other so-called 'flanking areas' such as consumer protection and health, he claims that they simply involved a 're-classification of activities already conducted by QMV under Article 100' (1998, p.55). What he fails to explain is *how* and *why* environmental decision-making 'escaped' from Article 100 in the period since the SEA (cf. Chapter 4). Golub (1996, p.718) is more circumspect about Britain's motives for accepting the environmental provisions, but he maintains that the outcomes (particularly the adoption of the subsidiarity principle which requires decisions to be made at the lowest effective level: see Jordan and Jeppesen, 2000) maintained British sovereignty and succeeded in trimming the Commission's expansionist ambitions. The remainder of this chapter analyses these events from a departmental perspective in order to assess how far the interlinked processes of integration and Europeanization in Britain accorded with what the various parts of the British Government initially expected to emerge from the IGC.

Negotiating positions

Social and economic measures

Douglas Hurd, Major's Foreign Secretary, and Chris Patten, who had recently moved from the DoE to chair the Conservative Party, oversaw the core executive's preparations for the IGC. Both men recognized the need to win over the Germans, and, much to Mrs Thatcher's disgust (Thatcher, 1993, p.475), set about courting them assiduously. Interestingly, this shift at the level of 'high' politics followed in the wake of a shift to a more engaged stance at the level of 'low politics' (i.e., in departments such as MAFF and, more recently, the DoE: see Chapter 2). In his first foreign speech as Prime Minister, Major told an invited German audience in March 1991 that he wanted the UK 'to be where we belong . . . [at] the very heart of Europe' (Hogg and Hill, 1995, p.79). Major (1999, p.268) had, in effect, reached the same conclusion that the DoE had reached about three years before: namely that 'it was better to play by club rules' by working within rather than outside the 'charmed circle' of France and Germany.

Like Thatcher, Major would have preferred not have had an IGC at all (H. Young, 1998, p.427), which he considered 'a negotiation before its time' (Major, 1999, p.264). But like it or not, an IGC was imminent, so he set about trying to stitch together a negotiating stance that pacified the anti-Europe section of his Cabinet while preventing the emergence of a two-speed Europe with Britain in the slow lane. But when it came down to the hard bargaining, the British contribution was almost wholly negative (Blair, 1999, p.1), except in respect of the enforcement and implementation of EU policy, which suited Britain's traditional strengths (Chapter 1). Throughout the IGC, the Cabinet was often as, if not more, divided on the issue of Europe than the Conservative Party

(Major, 1999, p.271). These circumstances allowed Eurosceptics in the core executive, such as Michael Howard, to tie Major's hands on key issues such as the social charter, where business pressure was the most focused (1999, p.358). In most other issue areas, including the environment, societal pressure was considerably weaker verging on non-existence (Forster, 1998, p.353), allowing Major greater leeway to 'package deal' (i.e., 'cut slack' with other leaders).

In December 1990, Douglas Hurd outlined the governmental (i.e., cross-Whitehall) position to the British Parliament (HC Hansard, Vol. 182, 6 December 1990), which was to: maintain the intergovernmental method; improve compliance with existing EU rules; promote greater subsidiarity; and maintain the EP's powers at the same level. The Commission, on the other hand, sought greater QMV, an integrated Treaty structure and a much stronger role for the EU in foreign affairs (*EC Bulletin*, 2/1991, pp.73–82). The EP, though not formally a participant, called for QMV to be extended to most policy areas and (of course) much greater co-decision-making (Corbett, 1992). It was supported by the Germans, who took the most ambitious stance of any state with respect to extending QMV and enhancing the powers of the Parliament (Dinan, 1994, p.170).

Environmental measures

Several European Councils preceding the IGC had expressed a desire for stronger EU environmental powers (Wilkinson, 1990). Therefore, the *real* question confronting negotiators in the IGC-PU was not whether to extend the EU's role (because it had already grown substantially since the SEA), but how far legally to formalize this informal expansion in the Treaties. Importantly, the failure of the legal firewall between Articles 130 and 100 meant QMV – though far from the norm – was already the *de facto* decision rule for many environmental measures by the time negotiators sat down to define an agenda for the IGC.

Changes at the international level not only supported, but actually demanded, stronger environmental protection from the EU. In the UN, European states and the Commission were beginning to prepare for the 1992 international Earth Summit in Rio. Around this time, several European political leaders were engaged in a high-profile political competition with their American counterparts to set higher standards in areas such as ozone depletion and climate change. Meanwhile, within the EU, the political fortunes of many national Green parties (including Britain's) were at an all-time high, and Green MEPs were beginning to make their presence felt in the EP. The Martin I and II reports, adopted by the EP in mid-1990 as an input to the IGC process, included ambitious proposals to strengthen EU environmental competences (Official Journal (OJ) C96 17 April 1990; OJ C231/97, 17 September 1990). Then there were the national environmental pressure groups. Recall that the vast majority of national-level groups completely missed the opportunity to lobby the SEA IGC; they did not make the same mistake again (Long, 1998, p.115). A consortium of the three largest groups, namely the EEB, the Worldwide Fund for Nature (WWF) and FoE, published a very detailed list of demands (which even suggested how to word key clauses and annexes) entitled *Greening the Treaty* (WWF, FoE, EEB, 1990). These included the adoption of QMV across the board, citizens' rights to environmental quality, access to environmental information, and an 'environmental guarantee' allowing greener states to introduce higher national standards than the EU. In Britain, the IEEP tried to capture this debate in a document (Wilkinson, 1990), which EPEUR circulated within EPG, and to UKREP and the Cabinet Office's European Secretariat.

To conclude, process and history mattered. The informal political integration since the SEA was both a cause and an effect of the political demands for higher environmental standards: a cause (particularly in Britain) in that it energized previously latent national environmental parties and groups: an effect insofar as it was nourished by their activities. In other words, the relationship between European integration and Europeanization was both intimate and highly reciprocal.

Preparations in Whitehall

Long before the IGC-PU was announced in June 1990, the Cabinet Office initiated discussions with Whitehall departments. The really big issues, such as the single currency and social policy, were determined in the full Cabinet, but the less party political matters such as the environment were passed down to different Cabinet committees (Blair, 1999, p.32). The available evidence suggests that the Cabinet Office quickly identified the environment as a possible 'sacrifice issue'. According to the EPEUR official that coordinated the British input to the Environment Council:

> the extension of QMV was one area where Britain was prepared to make a concession in order to gain proposals in other fields. Environment simply wasn't seen as a big issue and something worth fighting for. The UK had a bigger agenda elsewhere. It had some really big things that were show stoppers and other things

like the environment that weren't showstoppers that it could make concessions on.

(Shaw, 2000)

The DoE had balked at the prospect of stronger EU powers prior to the SEA IGC, but it approached the Maastricht IGC with equanimity. This was primarily because, as an organization, the DoE was now much more attuned to what was going on in Europe and felt much more comfortable working at that level. Domestic political support for environmental issues had reached unprecedented levels and EPG was growing in size (Chapter 2). The then Head of the newly-created EPEUR in the EPG, John Plowman (2000), remembers that:

> It [the IGC] was very much driven from the centre of the Cabinet Office and by UKREP. This was an area in which Ministers thought about surrendering sovereignty, not least because we were actually rather successful in environmental policy making, beginning to shape events and having much more influence over how policy developed.

The MINIS returns filed by DoE officials in the period 1989–91 certainly testify to a growing feeling of self-confidence. For example, the then Head of Water Quality, Dinah Nichols, reported that the EU was 'increasingly becoming the standard-setting body for all aspects of water quality', and claimed the DoE was 'now much better placed to influence the future direction of Community . . . policy' (DoE, 1989, 2.03, pp.11–14). Chris Patten's 1989 promotion to SoSE had also helped to make the department more receptive to some of the Cabinet Office's suggestions during the pre-IGC scoping process. His 1990 environment White Paper, a whole chapter of which was devoted to 'Europe', insisted that EU policy must be 'vigorous and forward looking' (HM Government, 1990, p.37). It continued: 'In the environmental field . . . the [European] Community has brought added value to Britain, and we have brought added value to it . . . The "clout" that the Community carries in the wider world is greater than the sum of each member's influence' (1990, pp.36–7). The contrast with the DoE's position prior to the SEA could not have been more striking.

By the autumn of 1990, the Cabinet Office and UKREP had identified a number of environmental issues that could arise in the IGC. These were: subsidiarity; QMV; removing the Commission's right of initiative; stronger powers (including fines) to enforce EU rules; and co-decision-making with the EP. EPEUR initiated a discussion about these points within EPG and with Nigel Haigh's IEEP. These indicated that the DoE had no obvious incentive to oppose a stronger subsidiarity clause (for many environmental issues, the EU was the obvious level to work). They also revealed that EPG did not feel as threatened by the possible extension of QMV as it once had. Indeed, to the extent that the DoE wished to upload policy, it could be helpful:

> We realised that we did not need to worry about QMV when enlargement had changed the dynamics in the Environment Council and there were countries [e.g. Spain] a lot further down the rankings who would do the business for us if we had to say 'no'. So people were more relaxed about it. The environment was one area of policy where we didn't need to insist on unanimity.
>
> (Plowman, 2000)

Nonetheless, EPEUR warned that more QMV could generate further misfits in politically important areas such as electricity privatization and the exploitation of North Sea oil (see Chapter 8).

Other departments were keen to curb the Commission's right of initiative, but EPG was worried that this could generate more misfits to the extent that the EP or other, greener, states proposed more ambitious legislation than DG Environment. It is a mark of how far the DoE's relationship with the EU had evolved since the SEA that the only issue on which it genuinely felt vulnerable was non-compliance. Here, the initiative came largely from the DTI, which had always suspected that Britain's 'dutiful' implementation record (see Chapter 2) was undermining the competitiveness of British industry (see DTI, 1993). This claim chimed with the core executive, which wanted to show the domestic electorate that it was fighting for British interests in Brussels. While not dissenting from these sentiments, the DoE predicted – correctly as it turned out (see Chapters 7 and 8) – that it would be one of the first departments to suffer if the governmental position won the day and the Commission was allowed to collect fines. These demands were fed into cross-Whitehall discussions, which resulted in the statement that Howe read out to MPs in December 1990 (see above).

To conclude, the DoE was not only strikingly more receptive to pro-European proposals than it had been in 1985–6, but it also had considerably more time and a more effective organization (EPEUR) to coordinate a departmental response to issues as they emerged in the IGC. However, there were very few issue areas in which the DoE actively proposed ideas either to the Cabinet Office or directly to the Commission prior to the IGC. For various reasons the two SoSEs, Patten and Heseltine, were unable to push the DoE's interests in Cabinet. Patten was at the helm during the orientation discussions, but was replaced by Heseltine just a month before the start of the IGC. Though a committed European, he spent much of his time trying to find a replacement for the controversial poll tax (Butler, Adonis and Travers, 1994, p.178 *et seq.*). He also had to tread carefully; his role in precipitating Thatcher's downfall had won him many enemies in Cabinet. It is indicative that a high-level meeting with Delors to discuss EPG's input to the IGC had to be cancelled when Heseltine was called back to London to deal with a leak about the abolition of the poll tax (ENDS, 194, p.3).

The Inter-governmental Conference

The Inter-governmental Conference commences

The IGC-MU covered a relatively tight agenda and was well organized by the Commission, together with a select group of finance ministers and central bankers. In stark contrast, the IGC-PU covered a huge sweep of concerns and was, as Delors remembers, a 'real nightmare' to orchestrate (Grant, 1994, p.181). To make matters worse, there was no scoping process akin to the Dooge Committee or the Reflection Group (see Chapter 6). Consequently, state representatives arrived at the negotiating table with a very mixed bag of demands and very unsure of one another's positions. A month into the IGC, there were well over 2,000 pages of submissions from different pressure groups (Corbett, 1992, p.276), and negotiations had become 'genuinely open-ended' (Forster, 1999, p.177). That anything emerged at all from this 'sprawling and heterogeneous' agenda (Pryce, 1994, p.39) was due in large part to the considerable patience and organization of the Dutch and Luxembourg Presidencies. For various reasons, the Commission was slow to come forward with its own proposals (Corbett, 1992, p.276) and, at key points in the IGC, found itself marginalized (Grant, 1994, p.181; Dinan, 1998, p.35).

The environmental dossier

As expected, most of the environmental changes were agreed fairly rapidly, appearing in the June 1991 'non paper' drawn up by the Luxembourg Presidency. This identified the contents of a possible 'draft treaty' (Luxembourg II: CSEC (91), 1295, pp.2–3). Of course, each nation arrived with its own 'pet' ideas – the Danish were particularly keen to amend Article 2, and

DG Environment reiterated its long-standing desire for stronger EPI (Wilkinson, 1990, p.4) – but the majority agreed to complete the legal codification of the environmental *acquis*. Apparently, the two most contentious environmental issues under discussion were the extension of co-decision-making (forcefully opposed by the British: see Ross, 1995, p.288) and the issue of financial support for poorer states (cohesion funding: see Corbett, 1992, p.290).

Mid-way through the IGC, the Court did what it had done during the SEA IGC and issued another landmark ruling on environmental policy. Although the ruling, which was issued just weeks before the publication of the Luxembourg II draft, did not radically alter the preferences of the main actors in the IGC-PU, it did emphasize the extent to which the *acquis* had evolved informally since the SEA. In the ruling on the regulation of the titanium dioxide industry (see Chapter 4), the ECJ tacitly supported the use of QMV and a greater role for the EP by arguing that the legislation in question should have been based on Article 100a rather than 130r. In so doing, the Court again confirmed that the proposals under discussion in the IGC were relatively benign (i.e., already part of everyday practice: cf. Chapter 4).

The DoE's direct contribution to the IGC agenda was, in contrast to the SEA IGC, positive but relatively indirect and circumspect. The first thing it did was push (via UKREP) for sustainable development, the leitmotif of the 1992 Earth Summit, to be given greater prominence in Article 2. This proposal probably originated in trilateral discussions between the EPEUR, Nigel Haigh's IEEP, and the big European environmental pressure groups. It was politically appealing to the core executive because it did not carry any obvious financial cost. It also did not involve the transfer of competence or undermine the core executive's espousal of greater subsidiarity. Its appearance in the final text suggests that these views were widely shared in the IGC.

The other suggestion was that DG Environment be allowed to assess the environmental impact of proposals made by other Directorates-General. For one reason or another this failed to make sufficient headway and was dropped from the final text. Interestingly, Heseltine claimed these two proposals were proof positive of a 'new determination . . . to wipe out the "Dirty Man of Europe" tag' (ENDS, 194, p.3). Heseltine's only other high-level intervention was towards the end of the IGC, when he wrote to the Dutch Presidency in late November 1991 to suggest a European inspectorate to audit the work of national environment inspectorates. Although this idea did not make the final text, it was eventually taken up in another form (IMPEL – the EU's Implementation Network of national officials), thereby confirming the DoE's ability increasingly to work with the grain of continental European thinking at the more technical level of 'small' decisions (ENDS, 201, p.3). In an interesting contrast with the SEA IGC, the only genuinely innovative proposal that the British core executive wholeheartedly supported in the IGC – namely, the application of fines for persistent non-compliance – troubled the DoE most. With hindsight, this was a very radical proposal, which stemmed from Whitehall's irritation at the inability of some EU states to implement policies. Having been at the sharp end of trying to resolve misfits between EU and British practice for a decade, the DoE knew full well that Britain had its own embarrassing implementation problems, namely bathing and drinking water! But all told, the DoE was, if anything, pushing a slightly *more* pro-integrationist line than many other departments and the core executive.

Having quickly secured agreement on the 'easier' topics, the negotiators moved on to the more controversial dossiers, such as foreign and social policy, and here the British contribution was much more negative. By the Spring of 1991, Britain was said to be 'systematically opposing every proposal for significant change at both IGCs' (Ross, 1995, p.154). Thereafter, the negotiations were disrupted when the Dutch, who held the Presidency, decided to bin most of the Luxembourg II draft in September 1990 (which it considered too unambitious) and start again, only to backtrack in the face of concerted opposition from almost every Member State (Corbett, 1992, p.291; Dinan, 1994, p.175; Ross, 1995, p.146; Major, 1999, p.270). However, a decision made by the Environment Commissioner, Ripa di Meana, to challenge Britain over the implementation of the EIA Directive (see Chapter 10) briefly threatened to disrupt the consensus on the environmental dossier. His intervention, which related to a number of controversial development schemes in the South East of England, was forcefully criticized by John Major. The media duly reported that Major had written to Delors warning him that Court action would make it harder for him to sign a Treaty (*The Economist*, 22 October 1990, p.42). The infringement proceedings were eventually terminated, but the episode reveals that the core executive was, by this stage, almost completely consumed with finding ways to demonstrate that it was 'acting tough' in the EU. Major referred to the spat in his showpiece speech to the Commons just a few weeks before the endgame at Maastricht:

> Majority voting was introduced . . . under the [SEA] and it could be extended under the new Treaty, but there must be limits to this

action. Whether a town by-pass goes to the east or to the west has nothing whatsoever to do with cross-frontier pollution or competition policy or any other aspect of the single market. These are issues that should rightly be settled at the national level . . . It makes sense therefore to codify and ring-fence Community competence . . . We must constrain the extension of competence to those areas where Community action makes more sense than national action or action on an intergovernmental basis.

> (HC Hansard, Vol. 199, 20 November 1991, cols 279–80)

It may only have been a coincidence, but land-use planning was one of just five areas of environmental decision-making which remained under unanimous voting following the Treaty.

Be that as it may, Major also told MPs that British negotiators had secured radical changes at a final conclave of foreign ministers held in Noordwijk in mid-November 1991 'that take us a long way from co-decision' (HC Hansard, Vol. 199, 20 November 1991, col. 279). What he actually should have said was that he had agreed to extend co-decision-making to a limited number of dossiers (including the environment: see Corbett, 1992, p.292; Pryce, 1994, p.50), in order to gain leverage in more important areas such as social and foreign policy, defence, and the use of the word 'federal' to describe the long-term goal of political integration (Laursen and Vanhoonacker, 1992, p.20). As in the SEA IGC, clearly environment was *not* a 'showstopper' issue as far as the British core executive was concerned, but something to be traded for bigger political prizes.

The endgame

The final negotiations at the Maastricht European Council (9–11 December 1991) were conducted by Major and a hand-picked team of officials from the core executive. Other Whitehall departments were not formally represented at the Council, although Major asked that the social security Minister, Michael Howard, and the head of the employers' federation, the Confederation for British Industry (CBI), be kept constantly in touch on issues concerning the social charter (Hogg and Hill, 1995, p.151). At the first *tour de table*, Major announced his concessions on the extension of co-decision-making and QMV in exchange for removing the 'f-word' (federalism) from the final text (Blair, 1999, p.82; Major, 1999, p.280). Social policy proved the most difficult dossier to agree, but Major eventually secured the opt-out that he needed to pacify his critics at home. Major duly claimed 'game, set and match' to Britain (a phrase he since disowns) (Hogg and Hill, 1995, p.157, fn.7), but the reality was a much more complicated mixture of trade-offs. Major gained opt-outs on monetary and social policy, but Britain remained distinctly at odds with the rest of Europe. Interestingly, the concessions he made in relation to the EP – the extension of competence and QMV to new areas, and the creation of a new role for the EU in foreign policy – went far beyond the very minimalist position advanced by Mrs Thatcher in 1990. The final outcome was much less ambitious than the EP had wanted, but even advocates of state-centric theories had to concede that it 'moved the EU very modestly in the federal direction favoured by Germany' (Moravcsik, 1998, p.449).

After the Inter-governmental Conference

Unlike the SEA, there was no sudden and unexpected burst of informal integration after the agreement of the new Treaty. In fact, the early 1990s marked a period of retrenchment and reform in areas such as water policy (see Chapter 7), with a shift towards less intrusive policy instruments such as voluntary agreements with industry. For once, political integration seemed to be moving in the more minimalist direction of which Britain approved. The deal did not, however, benefit Major politically, because he struggled for nearly a year to ratify the new Treaty in the teeth of determined opposition from backbench MPs (Baker, Gamble and Ludlam, 1993). As the public mood tilted sharply against deeper integration in the aftermath of the Danish 'no' vote, the new SoSE, Michael Howard, and his opposite number in France, pressed for the repatriation of certain environmental measures on the grounds of subsidiarity (e.g., troublesome water Directives: see Chapter 7).

In spite of widespread fears, the subsidiarity debate did not produce the fundamental 'rolling back' of EU environmental powers that many had predicted (S. Ward, 1997). Jordan, Brouwer and Noble (1999) show that the total output of new environmental policies slowed considerably in the period after Maastricht, with revised and updated legislation accounting for a much greater share than primary legislation. The Commission also trimmed its work programmes and withdrew a number of politically dormant proposals. However, it soon became apparent that environmental policy was one of the few areas in which the British and European public, energized by Europeanization, actively supported *greater*, rather than less, EU involvement. If anything, the subsidiarity debate powerfully revealed the robustness of the environmental *acquis*, which had, as process-based theories would expect, developed its own, powerful, internal 'dynamic'. So, when forced to decide what to 'roll back', the Member States and the Commission could not agree a common agenda, as amply demonstrated by the protracted process of revising some water rules (see Chapter 7). Finally, it is worth remembering that as the water directorate was pushing subsidiarity as a means to close existing misfits, other parts of EPG were hard at work uploading proposals (e.g., integrated pollution control) that eventually created new ones (Chapter 9).

Theoretical reflection

On the face of it, our empirical findings bear out the predictions made by state centric theories. The British government appeared to enter the IGC with a clear set of political and economic preferences, negotiated right across Whitehall. As Thatcher had done before him, Major 'cut slack' by excluding Parliament from the negotiations until it was absolutely necessary to do so. The outcome might have been far from ideal, but from Major's perspective it was far better than the original Franco–German plan for deeper monetary and political union. A detailed study of the negotiation of the environmental dossier also affirms Moravcsik's claim that supranational entrepreneurship during the IGC-PU was largely ineffective, with the Commission maintaining a notably low profile. As far as Britain was concerned, 'the state' appeared to function as a single unit and there was no 'home running', at least not by the DoE. The DoE might have entered the IGC-PU in a more positive frame of mind than the previous IGC, but it was largely excluded from the locus of policy development. If anything, the DoE's behaviour was more passive than one would have expected of a more European department (see Chapter 2), led by a political heavyweight (Michael Heseltine) strongly committed to advancing European environmental concerns. If the DoE was getting better at using the Environment Council to 'home run', it did not carry through into the IGC, where its role, and hence its influence was relatively weak. Finally, Maastricht did not unleash as many unintended consequences as the SEA (although the social protocol and EMU plagued the remainder of the Major government). Neither did it generate many unintended consequences in the environmental realm. The IGC may have failed to set limits on the future (and mostly informal) extension of EU competence, but it did clarify a number of ambiguities and, by giving prominence to the principle of subsidiarity, facilitated a probing reassessment of the whole policy sector. It is for these and many other reasons that state-centric theorists such as Golub and Moravcsik dismiss the SEA as an exception to the normal (i.e., state-dominated) pattern of events in the EU. This interpretation is extremely plausible. After all, British negotiators led by Major mastered the European agenda rather better than Mrs Thatcher, who was bounced unceremoniously into an IGC she did not want. However, it misses, or at least underplays the importance of, four aspects of the story. Crucially, these four aspects were not confined to the period immediately prior to and beyond the IGC.

First, process *did* matter by influencing the timing and the agenda of the IGC. By signing the SEA, states pre-committed themselves to holding another IGC, albeit on a much more restricted basis than before. When the time came, new political demands were stuffed on to the agenda, transforming what had originally been a fairly narrow review of foreign policy competences into something much more wide ranging. Throughout this process, preferences fluctuated greatly and there were several important forward linkages. Initially, Britain resisted another IGC, but it soon found initiatives (such as tougher measures to enforce EU law) to feed into the process when two were convened. Moreover, the IGC process culminated in an agreement to convene another IGC (which eventually led to the Amsterdam Treaty: see Chapter 6).

Second, Moravcsik (1998, p.470) himself concedes that the origin of state preferences was 'problematic' and 'difficult to disentangle' relative to the other big bangs. However, as others have suggested (Forster, 1998; 1999), Moravcsik underplays the party political origins of British preferences, which were in large part determined by political wrangling within Major's own Cabinet. By contrast, a process-based account suggests that British preferences were not determined exogenously (i.e., by societal actors operating within the territorial confines of the nation state), but arose *endogenously* as part of the continuous process of integration occurring in more technocratic institutional settings such as the Environment Council working groups. A snapshot of the IGC misses a crucial aspect of this process, which was the way in which the unintended consequences of the SEA fed into and informed the agenda of the Maastricht IGCs. In particular, Moravcsik greatly underplays the importance of the informal integration in the period 1985–90 in shaping choices. This left the DoE to choose between doing nothing and laying down clearer limits on the future extension of EU competence. This was a much narrower choice than one might have expected to find in a political system dominated by states. Ultimately the desire shown by state executives to codify what had, in effect, been pushed forward in other institutional venues (notably the Environment Council), coupled with more targeted pressure from supranational environmental pressure groups (policy feedback), produced a set of negotiating outcomes that cannot easily be reduced to the economic interests of independent sovereign states. There was, in short, far more forward momentum in the European political system than state-centric theories allow for.

There is also evidence to suggest that the informal integration prior to Maastricht also affected the preferences of particular parts of the state. In 1985, the DoE resisted European integration during the SEA IGC, only then to spend the rest of the decade trying to regain control as the number of misfits totted up. By 1989–90, roles had reversed somewhat as the DoE tried to 'think (and act) more European' while the core executive resisted deeper European integration to pacify critics in the Eurosceptical wing of the Conservative Party who had become more and more alarmed at the Europeanizing influence of the SEA (see Chapter 4). In a sense, Europeanization made the DoE more European, by forcing it to get better at closing the misfits between EU and national legislation.

This leads to a third point: there was no clear distinction between the 'high' politics of Treaty negotiation and the 'low' politics of daily policy-making; the two were reciprocally and recursively interconnected. State preferences were shaped and then reshaped via an endogenous process involving multiple feedbacks from earlier decisions. Even in the negotiations themselves, the Court's ruling on the implementation of an existing Directive (titanium dioxide) and the Commission's bungled attempt to implement the EIA Directive played their part in shaping the context of interstate negotiations. Indeed, the Commission's attempts to stop Britain from closing misfits by subverting Directives goes a long way towards explaining the DoE's opposition to Article 228 fines.

Finally, in viewing events over short time periods, state-centric theories do not normally measure the extent to which 'big' political decisions generate the outcomes expected of them. Clearly, the IGC-PU failed to generate a strong pulse of integration. Howard's campaign for greater subsidiarity produced some early effects and also some perverse outcomes (e.g., water policy: see Chapter 7) but before long, political integration re-emerged in the less politicized context of the Environment Council as subsidiarity metamorphosed into a much less radical process known as 'Better Lawmaking' (Jordan, 2000). Interestingly, the DoE emerged as an increasingly influential champion of new EU policies during the post-Maastricht period (see Chapter 3). Many of these new EU policy initiatives fitted Britain's preferred policy style, but some proved to be extremely costly (e.g., urban wastewater), some produced unintended consequences (e.g., IPPC: see Chapter 9) and all of them contributed to a continuing extension of the environmental *acquis*. In other words, the DoE began to advocate deeper European integration as a means of attenuating the Europeanization of domestic policy. Process-based theorists would regard this as policy feedback.

Finally, it is also worth remembering that the Maastricht Treaty marked the coming of age of the EP. Since then, MEPs, whose powers were further extended at Amsterdam in 1997 (see Chapter 6), have emerged as an important antidote to state control in the environmental sector, and a continuing challenge to the DoE's efforts to master Europe (Chapter 11). Time will tell whether Major's apparently 'minor' concession during the endgame at Maastricht produces the most unexpected impacts of any of the other formal changes made by the TEU.

6
The Negotiation of the Amsterdam Treaty

A little over two years after the Maastricht Treaty entered into force, the EU embarked upon yet another long and exhausting round of Treaty changes, which culminated in the signing of the Treaty of Amsterdam. The EU was mandated by Article N (2) of the Maastricht Treaty to undertake another IGC, but many Member States, not least Britain, considered it a 'conference too soon'. The FCO claimed that the EU needed a '3000 mile service', not another long period of upheaval (Langrish, 1998). The FCO did have a point: the complicated changes introduced by the greatly delayed Maastricht Treaty (Chapter 5)[1] had hardly had time to bed in, and the pain and anguish of the subsequent ratification process were still fresh in many people's minds. The three largest Member States were, in any case, preoccupied with domestic matters, and the Commission was still coming to terms with Danish 'no' vote to seize the initiative. In all probability, the Amsterdam IGC would not have taken place at all had the EU not been legally required to hold one. So, whereas previous IGCs had drawn upon a reservoir of political support for deeper integration, the Amsterdam IGC began for no other reason than that the EU was pre-programmed to renegotiate the Treaties.

This chapter explores how far the DoE shaped the environmental agenda in the run-up to the IGC as well as the Europeanization that flowed from the Treaty changes. It suggests that like the SEA and TEU IGCs, the Amsterdam IGC was primarily motivated not by environmental but geopolitical considerations. However, environment actors quickly seized the opportunity further to consolidate and embed the environmental *acquis*.

Amsterdam: an Inter-governmental Conference in search of a purpose?

The IGC was supposed to fill in the gaps left by the TEU, but the pressures introduced by the 1993 decision to enlarge the EU into Eastern Europe added greater urgency to the proceedings and an array of new political problems to grapple with. The agenda, which was supposed to have been fairly modest, quickly expanded to include such intractable issues as the transparency and accountability of decision-making, and the allocation of votes in an enlarged CoM. Quickly, this 'negotiation in search of a purpose' (Moravcsik and Nicolaidis, 1998, p.13) began to lose coherence and a clear sense of direction (F. Dehousse, 1999, p.64). More than any previous IGC, it was 'a melting pot of disparate measures lacking a coherent vision of either substantive co-operation in a given area or the future institutional structure of the EU' (Moravcsik and Nicolaidis, 1998, p.14). In the past, the EU would have looked to the Commission to steer the agenda but, after the problems with the Maastricht Treaty, its new President (Jacques Santer), decided to maintain a very low profile, particularly during the endgame (Dinan, 1997).

The Member States were so consumed by these 'macro' political issues that the environment nearly did not feature at all in spite of early and intense lobbying by environmental groups. The Commission's high command was preoccupied with other matters, Ritt Bjerregaard never really imposed herself on DG Environment during her tenure as environment Commissioner, and none of the largest Member States listed environment as a priority in their preliminary submissions. The environment might have vanished without trace had it not received a very late push from the new Member States of Sweden, Finland, Austria plus Denmark (ENDS, 255, pp.40–2).

So, what did the new Treaty produce? The Amsterdam Treaty, which entered into force in May 1999, introduced several changes to EU environmental policy: it made sustainable development a fundamental objective of the EU enshrined in Article 2 and the Preamble; it strengthened the EU's commitment to achieving EPI; it greatly increased the EP's role by introducing co-decision-making for measures adopted under 130r (or Article 175 under the new numbering system: see Chapter 1); and it provided an opportunity for greener states to introduce national standards that are stricter than EU rules (the 'environmental guarantee': see Jordan, 1998a; Pootschi, 1998).

Political antecedents

Amsterdam was supposed to tidy up the unfinished business left over from Maastricht, but there was little appetite for spring cleaning in the mid-1990s. The EU was undergoing one of its periodic bouts of deep introspection; politicians were in an especially sombre mood following the painful and very drawn-out process of getting the Maastricht Treaty ratified. Of those closely associated with the single market project in the mid-1980s (Chapter 4), only Germany's Helmut Kohl was still in power. But an IGC had to take place so, in June 1994, the European Council agreed to establish a Reflection Group of national and EU officials to prepare an agenda. There was a general consensus that the IGC-PU had been very badly prepared, and many hoped that this new arrangement would give the Amsterdam IGC a much sharper focus. However, nearly a year elapsed before the Group finally convened in June 1995. As with the TEU five years before (Chapter 5), states used this opportunity to look for related agenda items to throw into the ring. These included a whole raft of issues raised by enlargement such as the allocation of votes in the CoM and the number of Commissioners per Member State. Before long, pro-integrationists began to portray the upcoming IGC as a necessary step to enlargement rather than just a quick 'tidy up' (Dinan, 1999, p.170).

Whereas previous IGCs had been dominated by the Franco–German alliance, this time neither government wished to advance issues (e.g., harmonized asylum and immigration rules) that might upset their domestic electorates. The German Government's chief concern was how best to maintain the momentum behind monetary union. Although it made encouraging noises about extending QMV to new areas, it spent too much time looking anxiously over its shoulder at the reaction from the German Länder to drive the IGC forwards. Throughout large parts of the IGC, John Major's Conservative government (1990–7) remained implacably opposed to *any* move towards deeper integration. By early 1996, the Eurosceptics held the balance of power in the Cabinet, leaving Major with little option but to pursue a 'Euro-realist' position in the vain hope of holding his party together (Gowland and Turner, 2000, p.295) while preventing the formation of a 'two speed' Europe. To make matters worse, within months of the start of the IGC, Britain found itself embroiled in a fierce battle with the rest of the EU over controls to limit the spread of BSE. Under pressure from his own party and desperate to secure a lifting of the export ban on British beef, Major announced a short, and largely unsuccessful, policy of 'non-engagement' with the rest of Europe. A month later, in June 1996, Britain eventually fell back into line 'amidst much attempted face-saving' (Nugent, 1997, p.6), but by then the British had marked themselves out as the most awkward participant in the negotiations. Kohl publicly criticized Britain for being 'the slowest ship in the convoy' and warned that it should not be allowed to impede the rest (*The Independent*, 20 February 1996). Other states could also see the merits of a more flexible arrangement, which came to be known as 'flexibility' or 'variable geometry' in the negotiations (Edwards and Pijpers, 1997). The Labour Party sought to make political capital out of the situation and in early 1995 promised a more 'constructive engagement' with the rest of Europe (H. Young, 1998, p.485). In a headline speech, the Labour leader, Tony Blair, advocated the extension of QMV to social, environmental, industrial and regional policies. Other participants in the IGC soon realized that: 'the main rule of the game was to wait for the British election . . . Everybody knew that the IGC was heading for Tony Blair or crisis' (F. Dehousse, 1999, p.13).

The IGC itself lasted nearly 18 months and spanned four Presidencies. If we also add the six months of pre-negotiations in the Reflection Group, the whole IGC process extended over two years. It was also considerably more openly political than the TEU, energizing pressure groups that had not previously lobbied an IGC before (Mazey and Richardson, 1997, p.120). A public hearing organized by the EP in October 1995 (presumably in response to the post-Maastricht trauma about democratic acceptability) attracted well over 100 groups. Groups that had previously lobbied IGCs were also much better prepared and coordinated for Amsterdam (Stetter, 2001). Finally, Major's political problems meant that substantive negotiations were unlikely to begin until after the British election, thereby raising the political stakes in the final endgame.

The environmental provisions of the Amsterdam Treaty

The Amsterdam IGC followed the pattern of previous IGCs insofar as the environment did not play a prominent role in the build-up to, or during the negotiation of, the new Treaty. Nonetheless, the Amsterdam Treaty did make a number of important changes to the legal underpinning and functioning of EU environmental policy. These were as follows:

1 *Sustainable development*: a goal that environmentalists have fought for since the dawn of EU environmental policy, sustainability was elevated

to Article 2 of the new Treaty, where it functions as a fundamental objective of European integration.

2   *Environmental policy integration*: two new Articles (3 and 6) strengthened the EU's commitment to achieving EPI. Previously, EPI had been tucked away in the environmental part of the SEA and the TEU, and was somewhat narrower in its scope.

3   *The European Parliament*: co-decision-making was extended to areas of environmental policy previously subject to cooperation. With the exception of the five areas that remained under unanimity post-TEU (see Chapter 5), all environmental decision-making is now subject to co-decision-making.

4   *The 'Environmental Guarantee'*: new rules make it easier for individual states to go beyond harmonized EU standards where they do not afford a sufficient level of domestic protection.

5   *Subsidiarity, consultation and access to information*: other, more minor, changes clarified the meaning of subsidiarity, improved citizens' right of access to environmental information and empowered the Committee of the Regions.

The consensus is that these changes have consolidated the environmental *acquis* without adding any new principles, substantially deepening integration or extending QMV to new areas (Jordan, 1998a; Pootschi, 1998). As far as the environment was concerned, Amsterdam was indeed a 'tidying up' exercise. But although they may appear subtle, the long-term effect of some of the changes (especially the empowerment of the EP) remains unclear.

Extant accounts of the Amsterdam IGC divide neatly into two competing theoretical camps. State-centric accounts stress the importance of interstate bargaining and the weak input from supranational bodies such as the Commission (Moravcsik and Nicolaidis, 1998; 1999; Devuyst, 1999). Moravcsik and Nicolaidis (1999) attribute the 'modest' but 'uneven' outcomes to the unwillingness of most states to accept deeper integration. They admit that some issues were automatically included on the agenda because of pre-existing commitments, but claim that states never once lost control of the agenda. Admittedly, QMV was extended to 14 new areas, but these are dismissed as 'relatively uncontroversial' (1999, p.77) while the overall set of outcomes is said to correspond closely with the lowest common denominator of state preferences (1999, p.73).

In contrast, advocates of a process-based view of the EU draw attention to the underlying momentum of the integration process, which forced states into negotiating a new Treaty that few of them actually wanted. The institutional changes (such as the empowerment of the EP) were, in effect, a spillover from previous Treaty changes. Once the IGC was under way and the list of items for discussion began to lengthen, states found it more and more difficult to form consistent preferences or manage the agenda. In short, process and history once again mattered (Lord and Winn, 1998).

Existing accounts of the environmental provisions of the Amsterdam Treaty correspond to one of these two broad perspectives. Thus, Moravcsik and Nicolaidis (1998, pp.20, 24) dismiss them as 'modest', though they spend longer trying to account for the 'surprising' increase in the EP's power. By contrast, advocates of a process-based view highlight a number of inconsistencies in 'snapshot' explanations. For example, the explanation for the retention of unanimity in the five remaining areas of environmental policy is that they were 'extremely sensitive for one or (more generally a majority) of member states' (Moravcsik and Nicolaidis, 1999, p.77). Yet, this fails to explain precisely *why* and *how* they became sensitive. In the run-up to the SEA IGC, the environmental sphere was widely assumed to be politically and economically unimportant (see Chapter 4), but this changed as the *acquis* began to accelerate and deeply Europeanize national policies. Process-based theories point to the importance of the intervening processes of integration and Europeanization in altering the circumstances in which states make 'big' decisions. Germany's expectations about the operation of the environmental guarantee (which was originally introduced by the SEA) certainly proved to be wide of the mark (Moravcsik and Nicolaidis, 1998, p.23). The remainder of this chapter analyses the same events from a departmental perspective in order to assess the DoE's contribution to the IGC. It also investigates the extent to which the interlinked processes of integration and Europeanization post-Amsterdam have evolved in the manner predicted by different parts of the British Government.

Negotiating positions

Social and economic measures

When the IGC opened in Turin in March 1996, the Italian Presidency struggled to get discussions rolling because most Member States, while perhaps preferring a 'short and sharp' IGC, expected little substantive progress to occur until after the British election (McDonagh, 1998, p.54). There were no substantive negotiations during the Italian

Presidency but, even at this early stage, national negotiating positions were already very well formed (Griller *et al.*, 2000, pp.16–21, 29).[2] Thus, Germany's primary aim was to ensure that the IGC did not disrupt its plans for European Monetary Union (EMU), about which the German electorate was becoming increasingly mistrustful. France's main priorities were extending QMV, promoting enlargement and enhancing the EU's role in defence matters, whilst containing the powers of the EP. The British remained marginalized and weak throughout most of the IGC, so much of the early agenda was set by other states. The Conservative government's initial demands were laid out in a White Paper (published March 1996) entitled *A Partnership of Nations*. These included: no further extension of QMV; new limits on the capacity for informal integration; preservation of the existing three-pillar structure; a stronger application of the subsidiarity principle; containment of the EP's powers; and reform of the ECJ (FCO, 1996). No mention at all was made of the environment. Britain's barely disguised game plan was to turn the IGC 'into a non-IGC, that is a conference that will not make far-reaching changes to the Treaties', while diluting deeper integration by extending the EU into Eastern Europe (George, 1996, p.52).

Turning to the EU institutions, after delicate interstate negotiations, the EP was eventually granted an unprecedented, though still essentially minor, role in the IGC in order to give the whole process greater democratic legitimacy (McDonagh, 1998, p.59). Like the Commission, its preferences were also at the more maximal end of those fed into the IGC, encompassing an extension of QMV, greater co-decision-making (naturally!), more transparency and a strengthening of citizens' rights (Corbett, 1997, p.38). The Commission was much less assertive than it had been in previous IGCs. It had performed badly in the IGC-PU (see Chapter 5) and, in an attempt to regain the respect of states and European electorates, decided to pursue a much more 'pragmatic' line in the Amsterdam IGC (Dinan, 1998, p.37; McDonagh, 1998, p.209). Whereas Delors had been ambitious and intrusive, the new President, Jacques Santer, was pragmatic and discrete. In its formal submission the Commission called for an 'IGC with real ambitions . . . a genuine debate on Europe, on what Europe is about, on what course it is to take in the years ahead, and on everybody's role in this Europe' (CoM (96) 90, conclusion). Substantively speaking, this meant little more than a simplification of decision rules (i.e., the adoption of QMV across the board), a streamlining of Parliamentary decision-making procedures and new competences in areas such as employment (Corbett, 1997, p.37), in readiness for enlargement.

Environmental measures

Britain, France and Germany may not have agreed about much in the run-up to the IGC, but they were apparently united in their unwillingness to tinker with the environmental provisions of the TEU. A senior FCO official, Michael Jay, told a May 1995 meeting of British environmental groups hosted by the Green Alliance (a British non-governmental organization, or NGO) and the IEEP that 'there hadn't been much sign of interest yet' in pre-IGC negotiations (see Green Alliance, 1995). Just two weeks before the opening of the IGC in March 1996, there were still doubts about whether the environment would be discussed at all, as none of the big three states had listed it in their pre-IGC declarations (ENDS, 255, p.41). One of the participants in the IGC confirms that first-pillar issues such as environment and public health only made it on to the IGC agenda as the negotiations took shape, and then only because of pressure from the smaller (and in the case of the environment), greener Member States (F. Dehousse, 1999, p.36).

The Member States may not have been particularly interested in reforming the environmental *acquis*, but a wide variety of non-state actors were. In November 1995, a month before the publication of the Reflection Group's final report, a coalition of six major European environmental pressure groups published a detailed list of proposals entitled *Greening the Treaty II* (Climate Network Europe, *et al.*, 1995). They included the incorporation of the term 'sustainable development' into the opening articles of the Treaty, a stronger commitment to EPI, a citizen's right to environmental quality, across the board QMV allied to co-decision-making, a clear and unambiguous environmental guarantee, greater transparency, and an extension of the Commission's right of initiative to the Parliament. The EP was quick to produce similarly ambitious environmental proposals (Environment Watch: Western Europe, 1995a/b), but the relatively weak Commission was hesitant (see above), and did not produce a specific list of demands until almost a year later (CoM (96) 90). DG Environment officials admitted that it had been a struggle to persuade other, 'initially reluctant' DGs to find room for the environment in the Commission's submission (Environment Watch: Western Europe, 1996). Ritt Bjerregaard was not a particularly strong environment commissioner, and most of the Commission's position paper appears to have been drafted centrally in the Commission. In the end, the Commission's environmental demands were strikingly similar to the ones submitted by environmental NGOs and the EP, and included across the board QMV linked to co-decision-making, references to EPI in each and every chapter of the Treaty, citizens' rights to environmental quality and stronger enforcement powers (Calster and Deketelaere, 1998).

The Reflection Group

According to one participant, the Reflection Group was neither 'a reflection' nor 'a group', but an opportunity for national representatives to rehearse their pre-prepared negotiating positions (F. Dehousse, 1999, pp.4–7). Although the group included representatives from the EP and the Commission, the discussions were steered by foreign ministers (1999, p.94) rather than line departments like the DoE. The three largest Member States were unenthusiastic about reforming the environmental sections of the Treaty but, at a very late stage, representatives from the new Member States, namely Austria, Sweden and Finland (who had no previous experience of an IGC), combined with the Danes to force them on to the IGC agenda (Environment Watch: Western Europe, 1995a/b). The Group's interim report identified EPI, a stronger environmental guarantee and greater co-decision-making as issues of concern. It also noted widespread support for an extension of QMV (Council of the European Communities, 1995).

Having reflected on the agenda for six months, the Reflection Group issued its report in December 1995. This concluded that negotiations should 'focus on necessary changes without embarking on a complete revision of the Treaty' (Griller *et al.*, 2000, p.14). It identified three main priorities: making Europe more relevant to its citizens; making the EU work better and preparing it for enlargement; and giving the EU a greater capacity for external action (2000, p.6). Although it did not identify national positions, it is safe to assume that most of the references to a minority of 'one member' or 'several members' opposing a particular proposal were inserted at Britain's request (McDonagh, 1998, p.41). The introduction of sustainable development and the strengthening of EPI both secured support, but the extension of QMV (with Britain opposing), the environmental guarantee, stronger enforcement powers and greater co-decision-making were, to varying degrees, much less popular (CoM, 1995). Even at this very early stage in the IGC process, some of the more radical environmental demands, namely citizens' rights to environmental quality (which only the Commission and the Swedes seemed to favour), references to animal rights and access to judicial review, had been quietly dropped from the IGC agenda (Pootschi, 1998).

Preparations in Whitehall

In theory, the DoE could have used the IGC process to push for stronger environmental powers in Whitehall. Since Maastricht, it had become much more supportive of action at the European level, using the Environment Council to push for stronger environmental standards in areas such as IPC (Chapter 9) and wastewater treatment (Chapter 7). Potential IGC agenda items, such as the extension of QMV and co-decision-making, were no longer as threatening as they had been prior to the Maastricht IGC. Amsterdam was therefore a useful test of the Department's new, supposedly more European credentials: 'trying to be positive about Amsterdam was really rather important from a departmental point of view' (Summerton, 1999).

In many respects, the DoE prepared very thoroughly for the IGC. Long before the start date of the IGC was fixed, EPEUR contracted the IEEP to explore ways of 'greening' the EU to achieve sustainable development (IEEP, 1995). The IEEP's report included proposals to connect the sustainability reference in Article 2 with the commitment to achieve EPI in Article 130. Along with the provision of a clear legal base for EU environmental policy, these were long-standing aims of the IEEP's Director, Nigel Haigh (Haigh, 1998), and Amsterdam provided an ideal opportunity to complete the work initiated by the SEA IGC. The early portents were certainly very good: EPI and sustainability fitted Britain's broad policy agenda for the IGC (see above), being both minimalist and pro-subsidiarity. Britain had a 'greening government' system up and running in Whitehall (Jordan, 2001a), which it could have uploaded to Brussels. And, crucially, neither was likely to cost the British Government much to implement or to shift competence to Brussels (at least, not in the short term).

However, in spite of these careful preparations, the environment failed to make it on to the cross-Whitehall (i.e., governmental) position, which was published in *A Partnership of Nations*. There were several reasons for this. First, the core executive's near-neurotic fear of greater QMV and co-decision-making (which reflected the political crises triggered by the unintended consequences of earlier decisions) effectively ruled out any change in the environmental area. Departments such as Trade and the Treasury were adamantly in favour of holding this position to secure a strong 'opt-out' from further EU social policies. Major, too, had his own party political reasons for presenting a Eurosceptical face to the rest of Europe. Although an extension of QMV into taxation matters would have allowed EPG's environmental economics division to experiment with environmental taxes, the DoE was not entirely supportive of deeper integration, having its own departmental reasons for clinging to unanimity in some of the other exempted areas (see Chapter 5). Water resources was one very obvious example. According to the then Head of the Water Quality Division:

> Clearly we were apprehensive about extending QMV because there were areas in which it was very useful to the department to have unanimity . . . water resources for example was an area where we had been in deep difficulty [the bathing and urban wastewater treatment Directives], basically because we had done our sums and other countries hadn't . . . and other people were finding out what they had signed up to . . . We did not want to give up this benefit of unanimity when it was an area of considerable difficulty for the UK.
>
> (Summerton, 1999)

Second, domestic political pressure for change was relatively weak. The Council for the Protection of Rural England (CPRE) and the Green Alliance had held a seminar in May 1995 to generate enthusiasm (see above), but were told by one of the British IGC representatives that most states were decidedly unenthusiastic about the environment (CPRE and Green Alliance, 1995). With Nigel Haigh's assistance, the CPRE and the Green Alliance helped publicize *Greening the Treaty*, the document published by European environmental pressure groups, by sending copies to MPs and government departments, but neither had the resources or the time to mount a vigorous domestic campaign. The House of Lords made no mention of the environment in its review of preparations for the IGC, simply repeating the core executive's complaint about it being a 'conference too soon' (HOLSCEC, 1995, p.63). Academics, NGOs and EPEUR officials met in March 1996 to try to maintain momentum, but by then 95 per cent of the British negotiating position was already decided (Haigh, 1998).

Third, there was not much enthusiasm within DG Environment for Treaty changes in the early stages of the IGC. Press reports suggested that the Commission was reluctant to renegotiate environmental rules when the political climate was so Eurosceptical, in case it backfired and jeopardized the gains made at Maastricht. Therefore, on the advice of some states, DG Environment opted to maintain a low profile, rather than push for significant new Treaty changes (Environment Watch: Western Europe, 1996). This sudden turn of events alarmed the environment departments in the greenest Member States such as Denmark (International Environmental Reporter, 1995), which mounted a last-ditch campaign to have the environment debated (see above). Britain eventually responded in the July of 1996 with a strongly minimalist set of environmental demands, which bore the imprint of the core executive's guiding hand. These included: opposition to any extension in co-decision-making or QMV; the automatic withdrawal of old Commission proposals; the introduction of 'sunset clauses' in new legislation triggering automatic reviews (a reflection, perhaps, of the problems experienced in trying to update the water *acquis*: see Chapter 7); more systematic consultation with national parliaments and business groups prior to the publication of new Commission proposals; a more systematic application of subsidiarity; a review of the environmental *acquis* with a view to deregulating it; and the setting of clear limits on the ECJ's ability to advance integration informally (HM Government, 1996). These demands were strikingly more minimal than what EPEUR/EPG had pushed for in intra-Whitehall IGC discussions. They were also significantly more minimal than what the DoE was pursuing in negotiations over much 'smaller' decisions such as IPPC (see above).

The Inter-governmental Conference

The Inter-governmental Conference commences

The IGC proper commenced in March 1996 under an Italian Presidency (see above). Thanks to the pressure exerted by greener states in the Reflection Group, the environment was formally on the agenda, though most of the early discussion centred on more political controversial issues such as QMV, the EP's role in the IGC, and Pillar II and III issues. During the second half of 1996, the Irish Presidency tried to identify areas of early agreement, placing the more controversial questions to one side to deal with later (Nugent, 1997).

Now that environment was on the agenda, states seemed more than willing to countenance some modest, mainly presentational, Treaty changes, though a minority (including Britain) staunchly defended the status quo (McDonagh, 1998, pp.87–8). Quite early on, the Irish Presidency proposed that the Treaty be amended to strengthen and interlink the references to sustainable development and EPI. Apparently there was little dissent and these two changes made it into the final text (Calster and Deketelaere, 1998, p.17; McDonagh, 1998, p.88; Pootschi, 1998, p.77).[3] However, QMV, co-decision-making and other sensitive topics were simply too political for states to reach agreement on, so they were deferred until the endgame (Council of the European Communities, 1996). The environmental guarantee was notably absent from the Irish draft. Having been 'painstakingly negotiated' at Maastricht, there was 'a wide measure of reluctance . . . to re-open it in any substantive way' (McDonagh, 1998, pp.88–9). However, concerted pressure – especially from the Scandinavian states – in the final weeks of the IGC eventually delivered an agreed text.

During the first few months of its Presidency, the Dutch decided to 'park' most of the Irish Draft Treaty (including most of the environmental parts) in order to concentrate on the most politically sensitive topics (1998, p.138). Negotiations proceeded very slowly until the election of the Labour Government in May 1997 (Griller *et al.*, 2000, p.31), which began to 'melt the ideological pack ice which froze Britain's relations with the EU for so long' (*The Guardian*, 9 May 1997, p.15). Tony Blair immediately made good his promise to extend QMV to some new areas and accept greater co-decision-making with the EP in exchange for British 'opt-outs' on European border controls and defence. According to the Irish representative, his arrival transformed the negotiations: '[suddenly] there was a growing sense that a deal at Amsterdam was now on' (McDonagh, 1998, p.184).

The endgame

By the time the Heads of State assembled in Amsterdam for the endgame in June 1997, the environmental provisions (including the environmental guarantee: *ENDS Daily*, 1997; Jordan, 1998a, p.230) had just about all been resolved, bar majority voting. At one stage, The Dutch Presidency's sudden, last-ditch gamble to extend QMV to all aspects of environmental policy seemed destined to succeed (ENDS, 269, p.45). However, at the very last minute, Kohl – with one eye firmly on the German electorate and the Länder – overruled his environment ministry and withdrew his support, which effectively left the proposal dead in the water (Moravcsik and Nicolaidis, 1998, p.23).

Not surprisingly, most of the endgame was dominated by non-environmental matters such as the weighting of votes in the CoM and the distribution of Commissioners among the Member States in an enlarged Europe. Crucially, these and other matters were deferred to another IGC (Nice), which concluded in late 2001 (Jordan and Fairbrass, 2002). However, agreement was reached on sensitive matters such as flexibility, co-decision-making, employment policy and various Pillar II and III issues (for details, see: Moravcsik and Nicolaidis, 1998; Griller *et al.*, 2000). At the time, these outcomes provoked widespread dismay and criticism because they failed to prepare the EU properly for enlargement. Dinan (1999, p.178) blames an overloaded agenda and exhaustion among the summiteers. However, state-centric theorists maintain that the outcomes were modest but entirely realistic, because when it came down to the wire, states were simply unwilling to countenance deeper integration: 'Far from being the last of its kind, the Amsterdam Treaty is the harbinger of the future . . . [It is] the beginning of a new phase of flexible, pragmatic constitution building in order to accommodate the diversity of a continent-wide policy' (Moravcsik and Nicolaidis, 1998, pp.34, 36).

Theoretical reflection

It is still too early to assess the full impacts of the Amsterdam Treaty, which did not enter fully into force until 1999. At first glance, the negotiation and the resulting outcomes appeared to correspond closely to the predictions made by state-centric theories of the EU. Britain entered the negotiations with a clear set of governmental preferences negotiated across Whitehall, which were then bargained over in almost exclusively state-dominated arenas. The smaller states were successful in ensuring that the summit made some changes to the environmental *acquis* but the bargained outcomes mostly reflected the lowest common denominator of state preferences. So, for example, changes were made to Articles 2, 3 and 6, but more radical demands, such as the wholesale extension of QMV and the creation of rights to environmental quality, were filtered out of the discussions at a very early stage, dashing the hopes of supranational actors such as the EP and European environmental groups.

The DoE took active steps to prepare itself for the IGC by commissioning research in areas (such as EPI) where Britain had good ideas and successful experience to upload to the EU. However, it remained on the margins of a process through which the British position was developed and then articulated at the European level. Politically speaking, it found itself out of step with an increasingly Eurosceptical core executive. Consequently, environment never even made it on to Britain's shortlist of demands, which reflected the core executive's anti-integration/pro-subsidiarity agenda. So, in these and many other respects, the British state did appear to function as a single, internally coherent unit, with little evidence of autonomous action by the DoE. The Amsterdam Treaty made a number of modest adjustments to the environmental *acquis*, but they arose through the combined efforts of *other* national (environmental) departments, not Britain's.

However, there are several aspects of the whole IGC process which are at variance with this image of near-total state control. First, if Europe is dominated by states, why did the EU embark upon a Treaty renegotiation that few states actually wanted? Despite what state-centric theorists suggest, the functional links between successive IGCs imply that European integration has its own inbuilt dynamic, which enjoys some independence from state control. Thus, by signing the SEA, states committed themselves to the Maastricht Treaty, which in turn led to Amsterdam, and so on. On each occasion environmental demands were successfully fed into the process, even though they did not of themselves trigger the Treaty renegotiations. What had originally been billed as a 'routine service' (Langrish, 1998, p.19), soon metamorphosed into something much larger. Thus, by the end of the IGC, only four articles of the Maastricht Treaty remained untouched (1998, p.4).

Second, state-centric accounts fail fully to comprehend the extent to which party political considerations shaped the British government's preferences. As with the Maastricht IGCs, national pressure group activity was very limited, and almost non-existent in the environmental sphere. The main channels used by environmental groups to articulate their interests were European, not national (Stetter, 2001). The British governmental position was really a desperate ploy by Major to hold the two wings of his divided party together, *not* the outcome of a calm and rational assessment of what was in Britain's national economic self-interest. Blair, meanwhile, desperately wanted to appear more positive than Major, but was deeply unsure of what the British electorate would actually accept. His concessions during the endgame at Amsterdam were just as politically (as opposed to economically) motivated as those made by Major at Maastricht five years before.

This leads to a third point: there was no clear distinction between the 'high' politics of Treaty negotiation and the 'low' politics of daily policy-making; the two were reciprocally and recursively interconnected. Several of Britain's political demands were certainly developed in response to unintended consequences that had emerged slowly and informally throughout the development of the environmental *acquis*: for example, the unwillingness to surrender the national veto over water resources and the eagerness to trim the powers of the Court were partly a reaction to the political problems created by water Directives negotiated in the 1970s and 1980s (see Chapter 7). Indeed, it could be conjectured that so much had been achieved via informal means since the 1980s that it was not really necessary at all to discuss formal Treaty changes at Amsterdam: the job of entrenching the environment in the Treaty had effectively already been completed through a sequence of 'small' decisions at the sectoral level.

Finally, if there is one lesson to be learnt from the historical evolution of EU environmental policy, it is that the real impact of institutional changes usually only reveals itself in the fullness of time. So, while the Amsterdam Treaty may not have stimulated a sudden spurt of integration (cf. the SEA), it has generated much slower moving currents of change. For example, the renewed commitment to EPI, which was slipped into the IGC at a relatively late stage, has already pushed the Council and the Commission to embark upon the ambitious task of overcoming long-standing contradictions in EU policy, such as those between environmental and 'non' environmental policy sectors (Jordan, 2001a; Jordan and Lenschow, 2000). The new Article 6 has also given new political impetus to the Commission's hitherto unproductive efforts to extend environmental policy into new policy domains via the integration principle. In addition, there is the further empowerment of the EP which stemmed from a very late concession made by the British Prime Minister (rather like Major before him). This has already altered the context in which domestic environmental departments negotiate European environmental rules (Chapter 11). In the course of time, it could even represent the most important environmental legacy of Amsterdam, to the extent that it further erodes state control over the day-to-day expansion of the environmental *acquis*.

7
Water Policy: The Drinking and Bathing Water Directives

Having examined the three 'big' decisions, we now move down a level to the 8 'small' decisions made in four areas of daily policy-making, namely water, biodiversity, air and environmental assessment. The 1972 Paris Summit identified water pollution as a pressing political priority in the aftermath of the Stockholm conference (see Chapter 2). In a flush of enthusiasm, several EEC states signed international conventions (e.g., Paris, 1974; the Rhine, 1976) to protect shared rivers and the North Sea. But the Commission and the EP were the most instrumental in giving these intergovernmental agreements legal and political teeth by embedding them in EU water policy (Bungarten, 1978, p.166). This chapter analyses the history of two of the oldest and most important items of the water *acquis*, namely the Bathing and Drinking Water Directives. Today, the 1975 Bathing Water Directive (76/160/EC) is one of *the* most well-known EU environmental policies, but also one of the most poorly implemented. Long before it was finally amended in 1998, the 1980 Drinking Water Directive (80/779/EEC) was widely regarded as being scientifically outdated. Many states, including Britain, had struggled unsuccessfully for years to comply with the Commission's demand for compliance to the letter of the law. Contrary to earlier expectations, both Directives misfitted massively with pre-existing national policies in Britain. Correcting these misfits has cost British consumers billions of pounds and generated huge political problems in Whitehall. This chapter focuses on the DoE's handling of the two Directives and the ensuing Europeanization of national policy.

Historical background

The two Directives were originally based on proposals uploaded by the French government under the standstill agreement (Wurzel, 2002).

With explicit standards, formal monitoring procedures and timetables of compliance, they reflected a very French (i.e., *dirigiste*) style of legislation that, post-subsidiarity, is now out of favour in the EU (Jordan, 2000). Almost as soon as they were adopted conflicts arose with the very pragmatic, flexible and localized approach to water pollution that had evolved piecemeal in Britain over the course of a century. The DoE's decision to download the two Directives in spite of the potential misfit with British practices continues to haunt it to this day. The DoE now accepts that it negotiated them rather poorly, though no state can be said to have implemented either Directive perfectly. Environment Ministers adopted them because the political and legal stakes at the time were a lot lower than they are now. British companies certainly did not take EU environmental policy all that seriously: the first water directives were apparently adopted 'with little technical input from [the water] industry' (Thairs, 1998, p.156).[1] Both the DoE and industry genuinely believed that EU Directives were vague statements of intent, whose tacit objective was to improve water quality in continental Member States. As a former Head of the Water Directorate put it:

> in the 1970s, the nature of the European project wasn't really understood by the DoE. Directives were negotiated one by one, at a fairly low level in the department. The European project was understood as a kind of aspiration and Directives were seen as general objectives or guidelines. They were certainly not seen as binding law.
>
> (Summerton, 1999)

Before long, though, the two Directives began to be transmogrified as first the Commission and then national environmental pressure groups demanded implementation strictly to the letter and timetable of the law. As the misfit with British practices grew, Britain found itself under mounting political and legal pressure from the EU to spend vast sums of money cleaning up water to achieve European standards and defend itself politically against charges of complacency. Overall, the private water companies will have invested well over £39 billion in new treatment works by 2006, compared to £11 billion in the 15 years before privatization in 1989 (Water Services Association, 1996). The EPG's inability to manage the multi-levelled politics of integration and Europeanization arising from the two Directives almost scuppered the DoE's flagship water privatization policy in the 1980s, generated a public furore over water prices and landed Britain in the ECJ on at least three separate occasions.

In terms of the Europeanization of national policy, the two Directives have helped completely to transform pre-existing policy styles and administrative structures (Jordan, 1998a) (see Table 1.1). In a word, British water policy has, contrary to the initial expectations of most British commentators, been almost entirely Europeanized, though through no conscious or deliberate policy of the DoE. Rather, at several important stages, the Water Directorate found itself running hard to catch up with what one former Head, Dinah Nichols, candidly described as 'a rapidly shifting policy framework' increasingly determined by the EU (DoE, 1990, No. 2.03, p.13).

Pre-existing British policy

In theory, none of this should have happened because in 1973 Britain had highly developed procedures for dealing with water pollution. However, they did not fit with the EU's policy plans. To make matters worse, the British clung to them when challenged by continental states with similar objectives but different approaches. Traditionally, British water law only ever set out the broad goals of policy, leaving the detailed aspects of implementation to front-line professionals. In the 1970s, water was a 'classic example' of a technocratic sector of public policy, dominated by engineers and scientists (Maloney and Richardson, 1995, p.6) with the DoE's Water Directorate at the core. Tony Fairclough (1999) remembers that:

> It was resistant to initiatives from Brussels. It had settled ways of dealing with things. It wasn't root and branch opposed to *everything* coming out of Brussels, but it tended to take a very cautious attitude. The result was that every limit value suggested in water Directives . . . was questioned to the nth degree.

The traditional method of dealing with sewage from coastal communities typified many of these national characteristics. With a long coastline and relatively strong tides, Britain preferred to emit its sewage into the sea along pipes (where it would break down naturally) rather than treating it on land prior to discharge. With the exception of discharges to inland waters, there were no national legal standards. Local bureaucrats set informal limits on the basis of visual inspections, but they lacked legal teeth and sewage accumulated on beaches at famous holiday resorts. However, the MHLG readily accepted the Public Health Laboratory Service's (PHLS) advice that the health risk of bathing in sewage-contaminated seawater was 'trivial' even on aesthetically unsatisfactory beaches (Jordan and Greenaway, 1998). Scientists in the Government's PHLS, such as Brendan Moore (1954; 1975, p.104), forcefully maintained that the behaviour of bacteria in seawater was so complicated as to render fixed, numerical standards a complete nonsense. But of course, without legal standards, there was little legislative pressure on local authorities to spend money on sea outfalls and little was done.

The story was much the same with drinking water. Britain had a long history of concern, but prior to the EU's involvement there were no mandatory standards. Legislation dating back to the nineteenth century relied on the notion of 'wholesomeness' of supply. Before the EU's involvement, there were no mandatory standards defining what constituted 'wholesome', though water suppliers endeavoured to conform to relevant World Health Organisation (WHO) standards. The British water policy community took it to mean 'clear, palatable and safe' (HOLSCEC, 1976a, p.4), but these terms were never defined in law. Public concern about the health effects of nitrate and lead in drinking water was expressed as early as the 1950s (Hill, Aaronovitch and Baldock, 1989; H. Ward, 1993), but it rarely penetrated the mainstream political agenda partly because of the control exercised by the policy community of engineers and scientists. The epicentre of the community was the mighty Water Directorate within the DoE, which in the 1970s dominated (and was, to a large extent, philosophically and bureaucratically separate from) the EPG. According to a former Under-Secretary:

> the water industry was terribly self satisfied on the basis that we're the only chaps in Europe who can do this, so everything we do is super, even though it didn't have the money to spend on improving things. There was a temptation to think that no other country could do anything and we were the tops. Well that led to a kind of cyclic blind arrogance, which must have had its impact on Ministers to a degree.
>
> (Semple, 1999)

The Directorate closely involved itself in the work of intergovernmental bodies such as the OECD and the WHO, and expected the EU to function along similar lines (i.e., as a disseminator of best practice rather than a legislator). However, the Commission had a very clear idea of where it wanted EU policy to go and it was not, as Michel Carpentier explained, likely to be based on ideas uploaded by Britain:

> [W]e certainly recognise the long experience of Britain in environmental matters . . . But . . . incredible as it may seem, the rest of the

Community do not necessarily agree on the absolute paramountcy of axioms which are close to the heart of British experts and officials. For example the 'continentals' tend to believe more in standards defined on the basis of best technical means and applied through mandatory instruments. They mistrust systems based on goodwill and voluntary compliance . . . They have serious doubts about the absorptive capacity of the environment.

(in Levitt, 1980, pp.93–4)

The Bathing Water Directive

The origins of European Union policy

There are three reasons for the Commission's foray into this somewhat parochial field of public policy. First, at France's urging, the First Environmental Action Programme committed the EU to establishing quality objectives for certain classes of water, including water for bathing. Second, bathing water was already an issue of debate among European scientists and engineers. Britain's representative at international meetings, Brendan Moore from the PHLS, voiced his deep opposition to the use of fixed standards: British waters, he said, were affected by much greater (and therefore more cleansing) tidal flows than those in the Mediterranean (Wurzel, 2002). The WHO failed to agree precise standards but flagged the need for legislative standards. Third, in 1973, the French government uploaded details of its 1964 bathing water law to the Commission under the terms of the Standstill Agreement (see Chapter 2; it planned to extend pre-existing national controls governing seawater areas to inland waters).

In 1974, the Commission convened a group of national experts to develop water quality standards for different uses including bathing. The Commission had originally planned to follow the French approach and submit two parallel proposals covering fresh and seawater, but at the last minute presented the working group with a crude hybrid of the two (Wurzel, 2002). The DoE's representative, Brendan Moore (1977, p.269) was deeply alarmed at this, and complained at the lack of prior scientific discussion. He also claimed that there was no published scientific evidence in Europe to suggest that bathing in polluted water posed a significant health risk.[2] This doubtless affirmed the Water Directorate's suspicion that the proposal was concerned more with improving *amenity* than protecting human health.

The Commission finally published proposals in March 1975, which tried to upgrade the common interest in Europe. A draft dated February

1975 explained that 'no provisions of the same scope and the same degree of technicality . . . already exist in . . . national legislation' (CoM (74) 2255 final, 3 February 1975, p.1).

The negotiation of EU policy

Not surprisingly, Britain was 'very critical' of the Commission's proposal when it was finally published in 1975 (Levitt, 1980, p.110). The DoE fully expected some misfits to appear. For instance, it told British MPs that the proposal contained ambiguous terms, such as 'bathing water', that would be difficult to apply in practice, and estimated the total cost of compliance to be around £100 million. However, it believed that those misfits requiring new investment could be delayed until public finances permitted new spending. Their Lordships, on the other hand, said the problem was not so much of misfits with British practice, but the weak scientific basis of the proposed standards, which were 'so ill-defined as to be virtually unenforceable' (HOLSCEC, 1975, p.5). They also reiterated the British water policy community's claim that 'restrictions appropriate to the organised beaches of Southern Europe and the Mediterranean coastline . . . may be quite inapplicable to the tidal regime around our coast' (1975, p.4). Other commentators attacked the proposal from other standpoints. Lord Diplock, a distinguished law Lord, claimed that a *European* bathing water policy was almost certainly *ultra vires* on a strict reading of the Treaty of Rome. The water industry was also deeply critical: the National Water Council (NWC)[3] described it as 'the least satisfactory . . . Directive . . . to emerge up to now' (Levitt, 1980, p.111). The *New Scientist* (Tinker, 1975, p.66), a popular science magazine, spoke for many when it described the Commission's proposal as 'extravagant and unnecessary', being better suited to the waters of the Mediterranean which tend to be 'so grossly contaminated' as to be a 'regular source of disease'. Europhile Ministers such as Dennis Howell articulated similar prejudices:

> the EEC is mainly on about . . . the quality of bathing water in the Mediterranean . . . which of course is far more inferior in quality and therefore more hazardous than in this country . . . There are parts of the Community where the seas are disgraceful [and] where we would fully support . . . a cleaning up of the water.
>
> (HOCSCSL, 1975, cols 328, 332)

The Water Directorate claimed that the proposal was simply not relevant to Britain because British waters posed no serious threat to health and, in any case, were not frequented at the level envisaged by the Commission. There could, it continued, be no 'misfit' because Britain was already complying with the spirit of the proposal. If anything, on this particular issue, EU proposals 'misfitted' with (intellectually superior) British practices, rather than the other way round. Nonetheless, British politicians remained deeply suspicious of the Commission's motives, and the May 1975 debate in the EP consisted entirely of a consideration of amendments tabled by British MEPs (Levitt, 1980, p.96).

Throughout 1975, the Water Directorate worked with UKREP to minimize any remaining misfits. Changes made in the Council working groups and the Committee of Permanent Representatives (COREPER) included extending the compliance deadline from 8 to 10 years and creating a Management Committee to keep the Directive regularly updated. Throughout the course of the negotiation, UKREP sought to impress upon the Commission the DoE's belief that the standards were based mainly on *amenity*, not health considerations. In effect, the DoE tried to domesticate the proposal by convincing itself that there was no threat to health, and hence no justification for building expensive, new treatment works (i.e., there was no 'misfit' with national practices). However, we shall see that the Commission never really accepted this argument.

'[A]lthough far from perfect', the final text was regarded by the Water Directorate as an 'acceptable base for implementation' (Levitt, 1980, p.110). On 8 December 1975, a little under nine months after being formally proposed, the Bathing Water Directive was adopted by unanimity (although it carries the number 76/160/EEC). Thus, in spite of deep opposition from the scientific establishment, Parliament, the water industry, large sections of the media, many British MEPs and eminent lawyers, Howell adopted the proposal (a slightly puzzling decision given the painful and protracted problems that were to follow).

Formal[4] compliance

The aim of the Directive is to protect the environment and public health by raising or maintaining the quality of bathing water over time. It prescribes a list of physical, chemical and microbiological parameters, with the values that must be adhered to. In practice, the most important standards against which compliance is judged are those for total and faecal coliforms (a type of bacteria found in sewage but also associated with non-human sources, such as decaying vegetation and sea birds). The Directive leaves Member States to take 'all necessary measures' to ensure that bathing waters in their jurisdiction comply with the standards. The deadline was set at December 1985. Speaking as long ago as 1979, a Commission official admitted that the standards were 'far from perfect' but 'would not stand for eternity' (Gameson, 1979, p.213). However, stand they have. One of the first tasks specified by the Directive was to identify waters used for bathing. After much delay, the DoE drew up a series of guidelines for the UK Regional Water Authorities (RWAs), but emphasized the importance of interpreting them narrowly 'to keep to a minimum any additional expenditure incurred' (DoE, 1979, para. 12). Recall that this was a time of great budgetary constraint in Britain (Chapter 2), and Treasury-imposed restrictions prevented the RWAs from investing in new sewage treatment facilities. Reluctant and unconvinced of the merits of the endeavour, it was hardly surprising that the DoE submitted a list of only 27 'bathing waters' to the Commission (23 in southern England, the rest on the east coast); no inland waters, and no waters in Scotland, Wales or Northern Ireland were included. Popular resorts such as Brighton, Eastbourne and Blackpool were not identified. Leaving aside geographical and climatic factors, the difference between Britain's performance and that of other states, who between them managed to identify 8,000 waters, was still extremely stark (see Table 7.1).

Monitoring at the 27 waters in 1979 revealed that eight failed to comply with EU standards. Compliance fluctuated around the 66 per cent mark until the mid-1980s, with many waters passing into and out of compliance from one year to the next, mainly as a result of natural perturbations (see Tables 7.2 and 7.3).

Table 7.1   Number of 'bathing waters' in each Member State, 1980

State	No. coastal bathing waters	No. inland bathing waters	Length of coastline (km)
Luxembourg	0	39	0
Belgium	15	41	99
The Netherlands	60	323	850
West Germany	9	85	1,050
Eire	6	0	2,500
Denmark	1,117	139	3,400
France	1,498	362	4,140
Italy	3,308	57	5,500
UK	27	0	9,840

Source: Based on RCEP (1984), p.91.

Table 7.2 Total number of UK 'bathing waters', 1979–99

	1979–86	1987	1988	1989	1990	1991	1992	1993	1994	1995	1996	1997	1998	1999
England & Wales	27	360	364	401	407	414	416	418	418	425	433	447	457	461
Scotland	0	23	23	23	23	23	23	23	23	23	23	23	23	58
Northern Ireland	0	14	16	16	16	16	16	16	16	16	16	16	16	16
Total	27	397	403	440	446	453	455	457	457	464	472	486	496	535

Sources: Jordan (1997); Environment Agency.

Table 7.3 Compliance with EU requirements in England and Wales, 1979–99

	1979–86	1987	1988	1989	1990	1991	1992	1993	1994	1995	1996	1997	1998	1999
Total designations	27	360	364	401	407	414	416	418	418	425	433	447	457	461
Total in compliance	c.18	197	267	304	318	312	328	332	345	379	386	397	413	422
% compliance	c.66	55	73.4	75.8	78.1	75.4	78.8	79.4	82.5	89.2	89.1	88.8	90.4	91.5

Sources: Jordan (1997); Environment Agency.

Practical implementation and policy impacts

During the 1980s, four factors helped to realize what until then had been a potential misfit between EU requirements and British practices. First, EU requirements started to 'harden' as the Commission set about achieving 'full' compliance. Crucially, the then Commissioner, Stanley Clinton-Davis, sought to boost compliance across the EU by making an example of Britain:

> I found it especially offensive that Britain, the country with the longest coastline, argued that the definition [of bathing water] that they themselves had agreed upon, was vague. [Its] tactics were to abuse the Commission, challenge the nature of the Directive and submit huge amounts of data when fingered.
>
> (Clinton-Davis, 1996)

Thus, in April 1986, a few months after the passing of the compliance deadline (December 1985), the Commission initiated infringement proceedings against Britain, focusing on the waters at Blackpool and Southport. Second, pressure for change also began to build steadily at home as British society adapted to the Directive (i.e., societal lock-in). In its tenth report, the RCEP mounted a strongly worded attack. Under a picture of scores of bathers enjoying the water on a sunny day at Blackpool beach, it remarked: 'The bathing water at Blackpool beach – not identified as a stretch of water where bathing is "traditionally practised by a large number of bathers" ' (RCEP, 1984, p.xi). Other Parliamentary committees made similar criticisms (Jordan and Greenaway, 1998). Third, EU policy began to create and nourish new forms of national politics (policy feedback). For instance, national pressure groups began to draw attention to the legal misfit between national and European practices. In 1986, Greenpeace used its ship, *Beluga*, to sample coastal bathing waters around the British coast, assess them against EU standards and submit the results to the Commission.

Finally, but perhaps most important of all, from the mid-1980s EU requirements came increasingly into conflict with the DoE's desire to privatize the RWAs. As the sale proceeded, it dawned on Ministers that the City would not support the sale unless the liabilities of the water authorities (which, of course, included the as yet unquantified cost of implementing EU rules) were made absolutely clear. The importance of this point was not, it seems, fully appreciated when the Water Directorate began drawing up plans for the sale. In the Water Directorate's very worst-case scenario, an appearance in the ECJ would seriously disrupt that most fickle of commodities, investor confidence. Everything possible was done to neuter this threat. A series of leaked Cabinet papers (*The Times*, 1 June 1990) reveal the interdepartmental frictions triggered by the growing misfit between EU requirements and British practices. In one, the Water Directorate freely admits that the Directive could no longer be interpreted purely as a health measure: 'in spite of the medical advice that there is generally no harm, it is hard to defend a situation where (raw] sewage is discharged near . . . beaches and where the waters do not meet the Directive's standards'. But the underlying problem troubling the Directorate was the potential impact on privatization: 'An agreement with the Commission must now be preferable to an action with all the attendant publicity and uncertain outcome. An action before the [ECJ] during the privatisation discussions or the flotation would have wide ranging national implications going well beyond Blackpool.'

The Commission, however, refused to broker 'an agreement'. In fact, it did everything it could to force the British to transform its practices by maintaining the misfit. The DoE responded by designating more waters and offering token amounts of extra finance to buy time. However, the addition of so many sub-standard waters, coupled with changes in the sampling regime demanded by the Commission, had the undesired effect of pushing the overall rate of compliance to an all-time low of just 56 per cent (see Table 7.3). In 1988, the Commission cranked up the political pressure on the DoE by issuing two further Reasoned Opinions relating to the waters at Blackpool and Southport (Geddes, 1994, p.127). With water privatization reaching a critical stage, the DoE was forced to perform a delicate balancing act as it tried simultaneously to satisfy environmental groups, water consumers, the new water companies and the Commission (Jordan, 1998c). This involved giving hundreds of substandard sewage works temporary derogations, writing off the debts of the old RWAs and pegging future water bills at a high level (Byatt, 1996).

These arrangements bought the DoE a temporary breathing space and infringement proceedings were delayed until after the flotation (Bache and McGillivray, 1997). However, they also sowed the seeds of yet more unintended consequences. First, persistent non-compliance forced politicians to turn their attention from the bathing waters to the functioning of the sewage treatment facilities that caused the pollution. Eventually, the Commission was asked by the Environment Council to develop a proposal for an Urban Wastewater Treatment (UWWT) Directive setting down minimum standards for wastewater collection, inland treatment and disposal. This proved to be a significant and potentially costly departure from the objective of the Bathing Water

Directive, which was to improve water standards while leaving states to determine the means of achieving them. Published in 1989, the Commission's UWWT Directive proposal immediately and massively 'misfitted' with Britain's preferred solution to its bathing water problems, the construction of longer outfall pipes.

Under intense fire from greener states, the then SoSE, Chris Patten, overruled his advisers and the water companies, and committed Britain to a costly programme of improvements, without, it seems, adequately briefing Cabinet in advance (*Guardian*, 24 February 1990).

Second, the spiralling cost of the coastal water improvement programme required to implement the Bathing and UWWT Directives soon spilled over into the realm of party politics, where it became embroiled with an equally charged debate about the wider ramifications of the privatization process, particularly the benefits enjoyed by senior 'fat cat' executives and shareholders, and the respective roles of the regulatory watchdogs. Whereas before British water quality had resided within a tightly knit policy community of experts, by the mid-1990s it had become 'political poison' (*Financial Times*, 29 July 1994).

Finally, under legal pressure from the Commission, Britain has continued to designate more waters, including a number of inland sites (see Table 7.2). However, in spite of the huge sums of money spent, compliance continues to hover around the 90 per cent mark. The recognition that 100 per cent compliance is unachievable has accentuated the political pressure for reform (see below). Crucially, even when the most significant sewage discharges are removed (often at huge financial cost), natural perturbations can, as Moore rightly predicted, still shift a 'bathing water' into or out of compliance within hours. In 2001, the Commission reopened infringement proceedings against Britain, which could lead to another ECJ conviction and (following changes made at Maastricht) a daily fine of up to £70,000.

The Drinking Water Directive

The origins of European Union policy

The precise origins of this Directive are somewhat unclear. Drinking water was mentioned in the First Environmental Action Programme and national experts began discussing EU standards as early as 1973 (see above, and Bungarten, 1978, pp.203–5). However, no Member State appears to have consistently championed it. This may explain why consensus proved so much harder to achieve than in relation to bathing water, where there was an uploader of sorts (France). Part of the difficulty lay within the Commission itself: one part (Directorate-General V, social affairs) developed drinking water policy largely in isolation from another (the ECPS in Directorate-General III) that was working up proposals on surface water for drinking. The science of drinking water quality was also hugely complex and deeply contested, even in the early 1970s. Proposals for the Directive, which bore the strong imprint of French thinking, were not actually published by the Commission until July 1975 (CoM (75) 394), by which time the bathing water proposal was well on the way to being adopted.

The House of Commons discussed the proposal in December 1975. In a short debate, which centred on technicalities rather than whether the EU should even be involved in this most 'domestic' of policy areas, Howell suggested that a harmonization of standards would be: 'of tremendous importance to the people living in the Community . . . [I]t is of public importance to be able to guarantee the standard of tap water to British citizens abroad or to citizens of other European countries coming here' (HOCSCSL, 1975, p.7). Their Lordships scrutinized the proposals a lot more carefully and concluded that the compliance deadlines (of up to two years) were 'wholly unrealistic' (HOLSCEC, 1976a, paras 12–14) and warned of huge misfits, leading to serious implementation problems. They also noted that the Commission had proposed some standards (e.g., lead and nitrates) that 'misfitted' considerably with prevailing WHO and British standards. On that basis they confidently predicted that up to 10 per cent of homes in England and Wales would not conform to the lead standards, with an even higher percentage in Scotland.

The negotiation of European Union policy

That the COREPER discussed the proposal at over 60 meetings spanning a period of five years illustrates its political sensitivity (Demmke and Unfried, 2001, p.125). One of the problems was the sheer number of different chemical parameters involved. Ten Presidencies came and went in the five years, so at times the negotiators had no clear political direction. British representatives were particularly worried about the standards for lead and nitrate, while the Dutch pressed for very stringent standards for chlorides and conductivity (Haigh, 1992, section 4.4-3). In the end, the differences between national negotiating positions were successfully papered over not by altering the standards themselves, but by extending compliance deadlines, adjusting the sampling regime and allowing states to grant derogations in certain circumstances (though Britain worked hard to secure more lenient controls on lead concentrations). The need for unanimity in the Council also explains why the final text is so liberally sprinkled with imprecise terms and lacks clear guidance on important matters such as monitoring (Haigh, 1992, 4.4-4). Importantly, many of the precise numerical values attached to the parameters in the Commission's initial proposal survived virtually unscathed.

The Council finally reached agreement on the proposal in 1978, but it took another 18 months for the Directive to be formally adopted. Almost straight away doubts surfaced about the reliability of some of the standards, notably the parametric value for pesticide of just 0.1 µg/L. The problem was that, in the 1970s, not enough was known about the long-term effects of pesticides, so the EU set the standard at the minimum detection level (in effect, a surrogate 'zero').

Formal compliance

The Directive specified standards for drinking water, both directly and after processing, with the aim of protecting human health. Annex One listed no less than 62 physical, chemical and biological parameters, and the standards that must be achieved. The deadline for compliance was set at 1985, although derogations were permitted in certain circumstances. Two months after the transposition deadline, the DoE published a Circular transposing the Directive into British law. The presumption – at least within the Water Directorate – was that there were would be no serious misfits with existing practice: 'The importance of a wholesome water supply has long been recognised in the [UK] and the provisions of the directive accord with existing UK standards and practice. The directive will therefore *underline and reinforce, rather than alter*, existing policy' (DoE, 1982a, para. 2; emphasis added). The Circular continued: 'the [SoSE] will regard compliance with the directive as a necessary characteristic but not a complete definition of any water that is considered to be wholesome' (para. 5).

Behind the scenes, though, the Water Directorate was peddling hard to close potential misfits in relation to the lead and nitrate parameters. The 1982 Circular claimed that some of the maximum concentrations listed in the Directive should be regarded as absolute, whereas as others were just averages over time. In fact, the Directive made no such distinction, and the Commission started to take a closer interest. When threatened by the Commission with legal proceedings, the DoE prevaricated (Bomberg and Peterson, 1993). A week after the passing of the compliance deadline, the then Minister, Ian Gow (HC Debates, 23 July 1985, cok. 457–8), explained that the DoE would delay the misfit by *inter alia* granting derogations to supplies that breached EU standards. He also said that some of the standards were 'inappropriate' and asked the Commission to review the matter. The Commission, however, refused to close the misfit by weakening existing standards. It added that some of Gow's derogations were illegal because they had not been submitted early enough. It also pointed out that the nitrate standard related to each and every sample and not, as Britain claimed, to averages over a period of time. The Government eventually conceded this point and the derogations were formally withdrawn in 1988. In 1986, the Water Directorate informed the SoSE that: 'we shall have the hard task from now on of persuading the Commission of the need for flexibility in setting and administering concentration limits . . . The [EU's] drive to regulate . . . on a common basis remains an unhelpful and demanding diversion' (DoE, 1986c, 2.03, p.13).

Having received several detailed complaints from the British NGO FoE, the Commission initiated formal legal proceedings in 1987. By late 1987, the Water Directorate conceded that it could no longer minimize Europeanization by subverting EU legislation: '[the EU] has called for a much more quantified approach to drinking water quality than we have taken before, and has revealed various shortcomings. The increased attention to the enforcement of Directives . . . reinforces the need to take these statutory responsibilities much more seriously' (DoE, 1987, 2.04, p.12). The shift in position was undoubtedly accelerated by the exogenous pressure of water privatisation (see above). As with bathing water, the Directorate was terrified that the financial markets would balk at the cost of meeting EU standards unless they were properly quantified and accounted for in the pathfinder prospectuses. So, reluctantly, it conceded that every single sample of water, rather than just an average, would have to comply with EU standards. It also sought a speedy mechanism to adapt the Directive to new scientific information, but this was 'emphatically' rejected by Clinton-Davis, who said it would be 'very difficult politically' to achieve (ENDS, 166, p.14). Other states were, however, sympathetic to Britain's case, and high-level meetings of national government experts were convened in 1987–8. However, the Spanish government refused to countenance renegotiation, and the opportunity was lost (Jordan, 1999).

In the meantime, the DoE authorized hundreds of temporary derogations covering non-compliant sources to put the industry on a sound legal footing. These 'undertakings' (as they were known) must have satisfied the Commission because formal Court proceedings were delayed until the industry was safely in the private sector. The ECJ finally ruled against Britain in 1992, but by then almost all the breaches had been rectified. Since then, the work undertaken by the privatized water companies has greatly improved rates of compliance. By 1999, roughly 99.82 per cent of the 2.8 million water samples taken by British authorities met relevant EU standards (Drinking Water Inspectorate, 1999). However, in 1999, Britain once again found itself facing further infringement proceedings and the threat of daily fines.

Revising the water *acquis*

In the early 1990s, EU environment policy entered a period of retrenchment and reform. The main contributory factors included political pressure from water suppliers, improvements in scientific and technical understanding, and greater 'hands on' experience of implementation. Crucially, almost all states had begun to realize the full cost of achieving EU targets that they had rather hastily adopted in the 1970s. Clearly, this Directive was Europeanizing many states much more deeply than they had originally expected or desired. When two of the three largest states – France and Britain – drew up 'hit lists' of troublesome legislation in 1993, stories appeared in the media that both Directives were about to be repatriated to the national level.

Meanwhile, in other countries, the pressures to revise EU drinking water rules also built up in the late 1980s through more mundane processes of scientific investigation and ground-level implementation. This new information was codified in a set of new WHO guidelines, published in 1993, which gave added impetus to the Union of European Associations of Water Suppliers' (EUREAU) political campaign for amendment. EUREAU's proposals envisaged the removal of the universal pesticide limit in favour of standards based on the toxicity of individual substances. The British position that the DoE had negotiated across Whitehall would probably have remained confidential had a secret 'non-paper' not found its way to the press (Jordan, 1999). This revealed Britain's real demands, which also included individual pesticide limits. If adopted, these promised a 'very substantial deterioration' in existing standards (ENDS, 237, p.38).

In drawing up proposals (CoM (94) 612) for a new Directive, the Commission honoured its pledge to consult more widely. It also accepted the need to reduce the number of parameters, but proposed a significantly more stringent standard for lead that threatened to extend, rather than close, the existing misfit with national practices (HOLSCEC, 1996). Significantly, the controversial parametric value for pesticides was retained despite British pleadings. It soon became clear that many actors, including the EP, environmental groups, their Lordships, suppliers of treatment technology and, interestingly, water companies in some of the greener Member States, had significantly readjusted their activities around existing standards, locking them in place. British construction firms, who had profited handsomely from the construction of new treatment facilities, also threw their considerable political weight behind the 1980 standards (*Daily Telegraph*, 30 August 1993). After much discussion, a revised Directive was finally adopted in 1998 (98/83/EC). It includes a smaller number of parameters, but some of the parameter values are significantly tighter, and the controversial (and costly) pesticide parameter value remains in spite of British opposition.

The Commission's proposal for a new Bathing Water Directive (OJ C112, 22 April 1994) suffered a slightly different fate. The HOLSCEC (1994, p.24) alleged that certain aspects revealed 'a regrettable disregard of current science' and, in marked contrast to drinking water, faulted the Commission for not consulting widely enough. Following a short review during the second half of 1994, the EP's Environment Committee recommended that further discussions be halted until the Commission came forward with better parameters, a clearer scientific justification and better analytical methods. Thereafter, a succession of Presidencies (including Britain in 1998) politely ignored the Commission's proposal. In late 2000, the Commission finally withdrew it and set about drafting a new one. In January 2001, it issued revised plans which, if implemented, will slash current compliance rates, require another major round of investment, and a whole new emphasis on controlling agricultural pollution (CoM (2000) 860 final). Finally, 98/83/EC and the proposed new Bathing Water Directive will almost certainly require much new capital investment.[5] So far, the adaptation process has conspicuously failed to hold down the increase in water charges, which was the British Government's primary motivation for seeking revisions in the first place.

Theoretical reflection

Of the four areas of day-to-day policy covered in this book, water stands out as having been *the* most deeply Europeanized. In respect of the administrative structures, style and underlying philosophy of national water policy, pre-existing traditions have had to adjust to new, European policy frameworks and ideas. The paradigm underpinning British policy in these two areas, the tools used as well as the process through which those tools are calibrated, have all been transformed (see Table 11.1). Over the course of the last 25 years, Britain has been forced to implement

policies that civil servants and national scientific experts initially
regarded as being extremely costly, scientifically outdated and of dubi-
ous benefit to human health. The two Directives have also forced a radical re-ordering of national environmental priorities, elevating
compliance with EU rules above other important but nationally-derived
calls on finance, such as leakage reduction and inland water quality.
What does the saga – which is far from complete – reveal about the
utility of the two theoretical perspectives described in Chapter 3? On balance, the weight of the evidence supporting the process-based account
seems absolutely overwhelming. It is scarcely plausible to suggest that
the political outcome of these two Directives, which amounts to billions
of pounds in unplanned investment, fundamental changes in established principles and practice, severe political problems, the creation of
new agencies (i.e., the Environment Agency and the Drinking Water
Inspectorate) to oversee implementation, and the establishment of costly
monitoring and reporting procedures, are remotely consistent with
'state' preferences in the 1970s. In spite of unanimous voting in the
Environment Council and a very long history of poor implementation,
the alliance between the Commission and various national environmental groups *has*, contrary to Britain's wishes, succeeded in Europeanizing
British practices by creating (and sustaining) misfits with EU requirements. The deregulatory campaign waged by the three largest Member
States suggests that they had fundamentally lost control of the situation.
After all, if they were happy with the water *acquis*, why did they want to
change it?

Basically, a situation that the DoE's Water Directorate thought was
problematic – but essentially manageable – has gradually unfolded
along a series of unpredictable and unforeseen pathways. A snapshot
view of events surrounding the adoption of the two Directives simply
does not uncover the lapses in state control that occurred as implementation proceeded and new interconnections were made with fresh policy
initiatives such as the UWWT Directive. In the 1970s, Britain set out (as
predicted by LI) belatedly to domesticate the EU by subverting both
Directives, but soon came unstuck. The DoE did not purposefully
exploit international negotiations to secure domestic political objectives
in the manner claimed by LI, even though the 'slack' to have done so
was most almost certainly there (until privatization, British water policy
was quiescent and the RWAs relied upon the DoE to negotiate on their
behalf in Europe: i.e., to gatekeep). But the DoE failed to exploit this
golden opportunity because it misread the signs, ignored experts' warnings and hugely underestimated the force of EU law. Helen Wallace

would find evidence aplenty of 'irrationality, confusion and mistaken judgements' (Chapter 3) in this case study.

Since then, Whitehall departments have had to chase political circumstances as successive interventions by the EU have agitated and greatly politicized British water policy, locking the transformative effects of Europeanization in place. Moreover, Europeanization has, *contra* LI, generated a set of *new* domestic problems (e.g., the furore over water pricing; the protracted battle to privatize the water industry) rather than solving old ones. The idea that it has somehow 'strengthened' the state is simply not credible. Think of the failed attempt to reduce the escalation in water prices, for example, the desperate attempt to stall infringement proceedings to safeguard water privatization or the delicate inter-departmental negotiations to reconcile the privatization plan with EU environmental commitments.

Process almost certainly did matter and there were a number of significant policy feedbacks. Decisions taken in the 1970s opened up significant new opportunity structures for national and European politics, and circumscribed subsequent policy choices. The Bathing Water Directive also brought new and politically controversial issues on to the European agenda, such as UWWT standards, which then fed back into national politics, producing unintended consequences and constraining departmental autonomy. Throughout the whole saga, EU institutions were not simply instruments of state power; with the active support of domestic groups, they exerted an independent causal influence on the Europeanization of British policy.

Finally, the 25-year process has irrevocably altered the interests of many of the participants, not least the DoE's Water Directorate which has become much more environmental. For instance, the Directorate has started (somewhat belatedly) to use the discipline of EU law to secure higher environmental standards through the domestic water price fixing process against the wishes on the national water regulator, OFWAT. In the past, the Treasury kept a very tight lid on environmental investments. The EPG has also tried to improve Britain's reputation by being more proactive in Europe. For instance, Patten sanctioned ambitious new EU policies (e.g., the UWWT and Nitrates (1991) Directives) in the Environment Council which would not have been accepted in the same form by a British Cabinet.

8

Biodiversity Policy: The Birds and Habitats Directives

with Jenny Fairbrass

Of the four areas of secondary policy-making examined in this book, biodiversity (that is, the diversity of plant and animal life) stands out as being the most indirectly related to the single market. However, this did not stop the Commission from finding ways gradually to expand EU competence to Europeanize national practices. The Wild Birds (79/409/EEC) and Habitats Directives (92/43/EEC) are the two central planks of the biodiversity *acquis*. Together, they established a legally binding framework of protection. Before the EU's involvement, British policy had been mostly informal, voluntary and locally determined. In contrast to some aspects of pollution control, Britain justly regarded itself as a frontrunner in the development of national and international nature conservation policies. As a nation, the British care passionately about wildlife issues, especially animals. In 2001, the Royal Society for the Protection of Birds (RSPB) had over 1 million members, which is more than the combined membership of the three main political parties. Nevertheless, misfits arose between EU requirements and these long-standing national policies. With hindsight, Britain's long history of biodiversity protection did not make the job of adjusting to EU rules and philosophies any less painful or unpredictable. In many respects, the disadvantages of having innovated with national policies before the EU caught up made it considerably harder. This chapter examines the DoE's involvement in the development (integration) and implementation (Europeanization) of the biodiversity *acquis*.

Historical background

The EU's involvement in biodiversity protection had very modest beginnings. The concept of 'biodiversity' was, of course, not mentioned in the

Treaty of Rome (Chapter 2), but (unlike, say, water pollution) it also barely featured in the First Environmental Action Programme. This is not to say that there was no international biodiversity policy in the early 1970s (Member States had adopted measures as early as the 1950s[1]) but only that it relied upon intergovernmental cooperation, through agencies such as the UN. Like most other European states, Britain believed strongly that the legal competence to regulate the protection of biodiversity should reside first and foremost with national and sub-national agencies, on the grounds that they were better placed to tailor protection to local conditions and economic priorities. However, in the course of time, these international agreements created an important platform on which the Commission and other supranational agencies built an EU biodiversity policy.

Nowadays, EU biodiversity policy is considerably stronger and more intrusive than international biodiversity policy, although responsibility for setting the strategic policy framework, namely the 1992 Biodiversity Convention, is still coordinated by the UN. EU policy comprises a diverse array of different statutes (see Table 8.1). This chapter concentrates on the two most important Directives relating to wild birds and biodiversity. It reveals that these two EU measures have greatly disturbed many of the traditional features of British policy by placing environmental limits on the economic development of natural habitats

Table 8.1   EU biodiversity legislation

Subject	Legislation
Birds, animals, plants and habits	• Directive79/409/EEC: conservation of wild birds • Directive 92/43/EEC: conservation of natural habitats and forests • Regulation 3528/86: protection of forests against atmospheric pollution • Regulation 2158/92: protection of forests against fire
International agreements on the protection of species and their habitats	• Decision 82/461 – Convention on the conservation of migratory species of wild animals (Bonn) • Decision 93/626 – Convention on biodiversity • Decision 96/191 – Convention on the protection of the Alps • Decision 81/691 – Convention on the conservation of Antarctic marine living resources • Decision 98/493 – International Tropical Timber Agreement 1994
Seals	• Directives 83/129, 85/444 and 89/370: import of skins/products

(Freestone, 1996). As in the case of water and land-use planning policy (Chapter 10), the *acquis* did not expand in the predictable and containable manner that many Whitehall departments, including the DoE, had originally expected or desired. Although initially favourable to EU biodiversity proposals, Britain soon became wary of (and hostile to) further EU initiatives as integration and Europeanization accelerated beyond its grasp. The significant unintended consequences generated by the Europeanization of British biodiversity policy are still generating interdepartmental conflicts today.

Pre-existing British policy

Britain has always considered itself to be a pacesetter in indigenous biodiversity protection, having a history of legislation and political concern dating back to the nineteenth century (Reid, 1997, p.200). The size and political respectability of interest groups such as the RSPB (1889–), the Royal Society for the Prevention of Cruelty to Animals (RSPCA: 1824–) and the extensive network of local Wildlife Trusts (1912–), attest to the depth of the British public's concern for wildlife issues. However, these pioneering pieces of legislation were predominantly *species*- rather than habitat-centred. For instance, the 1872 Wild Birds Protection Act was primarily concerned with the protection of birds rather than the landscapes they inhabited. Throughout the twentieth century, many new laws were adopted, driven in large part by the efforts of species-oriented pressure groups such as the RSPB and the Fauna and Flora Preservation Society (1903). Independent expert committees were regularly convened to review the effectiveness of these policies, but most found no serious conflict between the goals of recreation, nature conservation and agricultural production (Winter, 1996, pp.193–8).

Since British officials believed these national policies had successfully stood the test of time and were superior to those in other countries, they tended to treat early EU initiatives with a certain degree of complacency (Lowe and Ward, 1998, p.9). What really pre-occupied senior officials in the CUEP (such as Martin Holdgate) were *international* wildlife issues such as tigers, whales and elephants, not indigenous biodiversity (see Chapter 2). These overseas concerns were mainly pursued through *international* channels (such as the UN Environment Programme) rather than European ones. Nevertheless, on joining the EU in 1973, the British seemed prepared to accept European biodiversity regulation, although they did little to initiate or sustain it.

Meanwhile at the national level, a tightly integrated agricultural policy community of interests, centring on MAFF and the National Farmers' Union (NFU), had dominated the management of the British countryside since the Second World War (Cox, Lowe and Winter, 1986, pp.183–4; M. Smith, 1993, pp.101–3). Environmental groups were almost entirely excluded from key decisions, as were powerful land owning interests such as the Country Landowners' Association (CLA), and even the DoE itself. Farmers had achieved this privileged position by dint of the post-war consensus on the need to intensify agricultural production to maximize national security. MAFF regarded the policy community as the rightful 'custodians' of the countryside, and considered that this community was on no account to be hindered by government-imposed nature conservation policies (Winter, 1996, p.200). Wherever and whenever conservation policies were adopted, they therefore embodied what is still known as 'the voluntary principle' (Waldegrave, 1985, p.109; Francis, 1994). This holds that government and landowners should work together to protect biodiversity. Thus, farmers can be encouraged (perhaps with financial inducements) to protect biodiversity, but never compelled by government fiat. The exemption of agriculture from the land-use planning controls introduced in 1947 not only confirmed but greatly strengthened the agricultural sector's immunity from environmental controls. New domestic wildlife policies were simply 'adapted to fit' these and other core policy priorities, such as improving transport links and exploiting new energy supplies (S.C. Young, 1995, p.238). For a long time, British farmers and their sponsors in MAFF successfully portrayed agriculture as a 'naturally conserving land use' (Lowe, 1992, p.5) that required little or no regulatory supervision. In practice, the massively subsidized agricultural sector has been one of the chief agents of extensive biodiversity loss and degradation since the war (Brown, 1993). The most recent national countryside survey (DETR, 2000a, pp.2–3) records the historical decline in the numbers and diversity of many bird and plant species, but adds that the 'negative trends in some key components of countryside quality . . . slowed or halted during the 1990s'.

To a large extent, this pattern of domestic power relationships was replicated at the European level. The British agricultural policy community centred on MAFF had integrated itself quietly into the much larger, though still very exclusive, European agriculture policy community, long before the Stockholm and Paris summits and the dawn of modern environmentalism. It was not only environmental pressure groups who found themselves excluded from the business conducted in the

Agriculture Council; DG Environment, MEPs and even national environmental departments often found themselves spectating from the sidelines as environment-damaging European policies were developed by the EU agriculture policy community.

The Birds Directive

The origins of European Union policy

As early as 1971, the EP called for wild birds to receive EU protection. These demands, which were stimulated by public disquiet about the (continuing) slaughter of migratory birds in Southern Europe, led indirectly to biodiversity's inclusion in the First Environmental Action Programme. Although it was heavily slanted towards pollution control, the Programme noted the need for European policies that 'ensure the sound management of . . . nature' and 'improve the environment in its broadest sense' (OJ C112, 20 December 1973, p.5, p.11). This could conceivably have included birds and habitats, though neither term appears in the published text. However, the Programme nonetheless succeeded in giving the Commission an all-important 'foot in the door' (Héritier, 1999, p.58), which it used to carve out a bigger niche for itself. However, aware that proceeding with undue haste might upset national sensitivities, at first the Commission confined itself to reminding Member States of their commitments under international law (Wils, 1994, p.219). A Recommendation to this effect was issued in December 1974.

Soon, though, the rising tide of public concern encouraged the ECPS to move a little faster. Particularly influential was 'a sustained campaign' by Dutch and German citizens directed at the Commission and MEPs (Gammell, 1987, p.2). During the mid-1970s, the RSPB in Britain identified European legislation as 'the top policy priority . . . [and] . . . substantial resources were thrown at it' (Pritchard, 2000). In late 1974, a transnational interest group called 'Save the Migratory Birds', consisting of the RSPB and other national and international animal protection organizations, presented a petition to the EP. A few months later, in early 1975, the EP responded by issuing a Resolution demanding EU regulation. According to the RSPB's campaign officer at the time: 'My suspicion is that the Commission probably wouldn't have taken [the Directive] forward of their own initiative – if it hadn't been for pressure from the NGO movement outside plus a willing, dedicated, enthusiastic MEP [Hemmo Muntingh]' (Hepburn, 2000).

The EP's involvement gave the Commission, which had been busily preparing the ground by conducting benchmarking studies and consulting national nature conservation experts, the green light, and it issued a proposal in 1976 that 'turned out to be more comprehensive than the Parliament had ever suggested' (Haigh, 1989a, p.293). As explained in Chapter 2, the Commission often tries to build up a scientific consensus on an issue to 'soften up' Member States in readiness for EU regulation (Héritier, 1999, p.59).

The negotiation of European Union policy

The British gave their approval to the proposal 'with little hesitation' (Haigh and Lanigan, 1995, p.22) because the DoE's Wildlife and Countryside Division believed that it fitted neatly with existing British policy and practice. However, any fit would have been entirely coincidental because the Department made no serious attempt to upload British law and expertise to the EU. Haigh (1989a) notes that, at the time, Britain had one of the most extensive bird protection regimes of any European country. Public opinion, even in rural areas, had always favoured bird protection, and the general expectation was that any misfits were more likely to appear in pro-hunting countries, such as France and Italy, than Britain. During the Parliamentary scrutiny process, the DoE was asked whether the Directive would affect the use of agricultural land. A civil servant responded:

> I do not see the directive making any great deal of difference in this respect . . . We would not see this directive as imposing any greater obligation in the UK than we are already doing . . . I do not think there is anything in this directive which tilts the balance [between conservation and agricultural interests].
>
> (HOCSCSL, 1977, pp.5–6)

Even environmental bodies such as the RSPB expected the 'misfit' between EU and national policy to be negligible, so the impacts of Europeanization would be 'either minor, can be coped with by derogation, or are advantageous' (1977, p.14). Under questioning, the RSPB's representative went as far as to suggest that: 'the main value of this Directive will be in improving the legal situation in . . . Member States . . . which at the moment have inferior bird protection laws to ourselves' (1977, p.14). During the subsequent Parliamentary debate (HC Debates, 1977–8, 17 November 1977, cols 937–40), the environment Minister said that: 'The Government have consistently expressed strong support for the principles of the Directive . . . [It] is likely to emerge in a form even closer to our present legislation and . . . require few amendments to the Protection of Birds Acts.' During the remainder of the debate, speaker after speaker extolled the virtues of British practice. One member of the scrutiny committee even remarked that 'we can pat ourselves on the back for our legislation and for the fact that it is a pattern for the Community' (HC Debates, 17 November 1977, col. 942).

The widely held assumption that Britain, while not an 'uploader' of national standards, was comfortably in among the leading pack of European states was borne out when the French and the Italians emerged as the chief opponents of the Directive. Once their concerns had been accommodated, the proposal was adopted by the then Eurosceptical SoSE, Peter Shore, in the dying days of the last Labour Government (April 1979). Even at this very early stage, more farsighted commentators predicted that serious misfits would eventually appear. Rehhbinder and Stewart (1985, pp.214, 247, 262) said it marked a 'fundamental departure from long established customs in some Member States' and suggested that further EU involvement in biodiversity protection was highly 'improbable'. In spite of these warnings, the Birds Directive was already part of the *acquis*. Moreover, there is no reason to believe that the DoE's acceptance of it was part of a surreptitious plot to 'home run' against 'non' environmental departments such as MAFF. It is more likely that all departments naively assumed that the new Directive was a benign (i.e. non-disruptive), technical matter, which would not disrupt the main aspects of British policy.

Formal compliance

In comparison to other environmental directives, the Birds Directive places extraordinarily broad duties on Member States, such as the maintenance of wild bird populations and a sufficient diversity of habitats, before going on to create a series of more specific obligations. These are to be achieved by establishing a network of special protection sites or areas (SPAs). The Directive was transposed into British law via the Wildlife and Countryside Act (1981), the Parliamentary passage of which provoked a fierce conflict between the British agricultural policy community and nature conservationists. The Act has been described as a 'logistical triumph' for farmers and other land-owning interests (Winter, 1996, p.206) because it embodied the voluntary principle (Adams 1986, pp.93–111). Among other things, the Act made Sites of Special Scientific Interest (SSSIs) a legal linchpin of the national nature conservation policy. However, the protection offered by SSSI designation has never been absolute given the primacy of voluntary action (see above). In 1981, approximately 4 per cent were said to be suffering serious damage, primarily from agricultural operations (ENDS, 68, p.6). The 'voluntary' philosophy also made the task of notifying new SSSIs extremely time consuming. Because the DoE saw SSSIs as a superior means of protecting birds, it also had the knock-on effect of retarding the implementation of the Directive. By July 1982, some 15 months after the formal transposition deadline, Britain had submitted a few candidate SPAs to the Commission (Pritchard, 1985, pp.2–3). In 1984, the Commission wrote to the British government requesting further information and eventually issued a Reasoned Opinion. In the view of one RSPB campaigner, Britain was guilty of:

> slow, minimalist listing of important sites; weak resolve in protecting habitats; excessive and inadequately controlled killing of birds; Government secrecy; and inadequate consultation . . . While comparing favourably with some Member States, Britain's implementation of the Directive falls short of the international example it could set if Government applied its provisions in a more positive way.
>
> (Pritchard, 1985, abstract)

One of the key problems was the 'voluntary' character of existing British nature conservation. Thus, in order to establish an SSSI, the government's nature protection agency, the NCC, first had to negotiate a Voluntary Management Agreement with the owners of the land concerned.[2] The process proved to be much more time consuming than the DoE had originally anticipated. These delays were exacerbated by the attitude of the then SoSE, Nicholas Ridley. A confirmed Eurosceptic, a hunter and a pro-farmer, he instructed his officials to 'make haste slowly' (Sharp, 1999). This placated cognate departments such as MAFF and the DTI, who resisted any external (i.e., EU) restriction on the economic development of land and Britain's coastal waters. According to Robin Sharp (1999):

> One of the main reasons why [the DoE was] so slow to actually designate the sites under the Birds Directive was the hostility of other departments to actually implement it because certainly once you have done it, it does limit the activities that can go on . . . and may lead to infraction proceedings.

By 1988, Britain had only managed to identify 32 SPAs (see Table 8.2), although the NCC claimed to have identified a further 188 'candidate'

Table 8.2   Numbers of protected areas in the UK, 1984–97

Year	Marine nature reserves	Special Protection Areas	Ramsar/wetland sites	Sites of Special Scientific Interest
1984		7	19	4,225
1985		7	19	4,497
1986		18	29	4,842
1987		22	31	4,724
1988		26	35	4,996
1989		33	40	5,184
1990		33	40	5,435
1991		40	44	5,671
1992		48	53	5,852
1993	2	77	70	5,964
1994	2	97	82	6,103
1995	2	104	88	6,178
1996	3	126	100	6,181
1997	3	150	111	6,264

Source: DoE, Digest of Environmental Protection Statistics (various years).

SPAs (HOLSCEC, 1989, p.9). In public, the DoE blamed the delay on the misfit between European and (superior) British requirements:

> The rate at which SPAs have been classified has been slower than expected . . . [T]he [DoE] firmly believes that sites must be adequately protected under UK law before being classified as SPAs. . . . Unfortunately, the considerable work involved . . . has proved more demanding on staff resources and time consuming than expected.
>
> (*ibid.*, col. 234)

Marine areas remained a particular problem (see Table 8.2). By 1989, Britain had only designated 40 Ramsar/wetland sites. This convinced some critics that the real aim of the exercise was to maintain sufficient leeway for future economic development, rather than provide biodiversity with absolute protection (HOLSCEC, 1989, p.10). Admittedly, Britain was not the alone in experiencing problems; according to the European Parliament (1988), implementation was 'defective' in most member states. Nevertheless, in comparison to states of a similar size, Britain has performed relatively poorly (see Table 8.3).

Practical implementation and policy impacts

In the early 1980s, DG Environment began to step up enforcement by taking cases of suspected non-compliance to the ECJ. Several of these

Table 8.3   SPAs notified to the Commission (as of November 2000)

Member State	Number of sites classified	Total classified area (km²)	% of national territory
Belgium	36	4,313	14.1
Denmark	111	9,601	22.3
Germany	617	21,672	6.1
Greece	52	4,965	3.8
Spain	260	53,602	10.6
France	117	8,193	1.5
Ireland	109	2,236	3.2
Italy	342	13,707	4.6
Luxembourg	13	160	6.2
The Netherlands	79	10,000	24.1
Austria	83	12,080	14.4
Portugal	47	8,468	9.2
Finland	451	27,500	8.1
Sweden	394	24,647	5.5
UK	209	8,648	3.5
Eur15	2920	209,792	

Source: http:// www.europa.eu.int/comm/ environment/news/natura/index_en.htm

drew upon information submitted by national environmental groups, which had consciously adopted a 'watchdog' role (Pritchard, 2000). Two of the most important cases were C-57/89 (*Commission* v. *Germany* (with Britain intervening), also known as Leybucht Dykes)[3] and C-355/90 (*Commission* v. *Spain*, also known as Marismas de Santona). The Court's decisions in these two cases appeared to elevate ecological considerations over economic concerns in the designation and management of SPAs (Ball, 1997, p.217; ENDS, 260, p.43). The DoE's lawyers had warned that this could happen, but in parts of EPG and in most other Whitehall departments the ECJ's rulings came as a complete shock. MAFF and the DTI had always enjoyed the right to trade environmental priorities off against economic and social concerns, and both wanted to keep it that way. Very few people, least of all in Whitehall, had ever expected the Directive to intrude so deeply into 'national' practices.

Apart from the Commission and the ECJ, the other main advocates of a more maximal interpretation of EU law have been environmental pressure groups. In the 1980s and 1990s, groups such as the RSPB, the WWF and BirdLife International became more and more adept at flagging cases of suspected non-compliance to the Commission and the media. In one case, these complaints led a senior Commission lawyer to pay an unprecedented 'fact finding' visit to Scotland. The very idea that the Commission might one day investigate British habitats against EU criteria would have been anathema to a Eurosceptic like Peter Shore, who originally sanctioned the EU's involvement.

Britain, too, has been caught by the case law stemming from the Birds Directive. In the first half of the 1990s, the RSPB challenged the British government in the British courts. The case, concerning Lappel Bank, was referred to the ECJ, which judged that the British government had been wrong to exclude an area of land from an SPA to allow development of a nearby port. Again, the ECJ supported a much more maximal interpretation of EU law than Britain had originally anticipated or desired. Since then, other cases have been settled informally between the Commission and the DoE. For example, the DoE staved off the threat of legal proceedings by promising to provide 'compensatory measures'[4] (ENDS, 260, p.43) for an area of mudflats affected by the Cardiff Bay barrage. These cases were not that financially burdensome, but they significantly constrained Whitehall's ability to determine national economic priorities.

The Habitats and Species Directive

The origins of European Union policy

The precise origins of the Habitats Directive are unclear. Some commentators trace it to commitments made in the Third Environmental Action Programme, others to the Fourth. Be that as it may, conservationists had long campaigned for additional EU policies to support the Birds Directive (Hatton, 2000). The Commission, too, always recognized that the Birds Directive did not offer sufficient legal protection to bird habitats, but knew there were political limits to what could be achieved in one legislative step. The main problem was that states were simply not prepared to countenance European habitats legislation in the 1970s. But a decade later, they were. The Commission can take a great deal of political credit for having facilitated this shift in state preferences. In 1980, a year after the adoption of the Birds Directive, it completed a benchmarking survey of national protected areas in collaboration with national authorities and international organizations (CoM (80) 222, p.34). The Commission had inserted a reference to this inventory in a draft Council Resolution, which was eventually adopted around the same time as the Birds Directive. However, it sensed that it was still not the right time for a *European* habitats policy. So, throughout the 1980s, it worked inside international organizations such as the UN to build the necessary political consensus, as well as strictly to enforce the Birds Directive (CEC, 1984, p.56; Johnson, 2000).

In the late 1980s, the politicization of environmental issues post-SEA (see Chapters 2 and 4) finally created a propitious political context and the Commission stepped up a gear (Dixon, 1998, p.224; Sharp, 1998, p.40). As with the Birds Directive, British and international environmental groups were highly supportive of greater European protection. An annual conference of British nature conservationists in the mid-1980s called for 'a missing law' to extend bird protection law to their habitats (Hepburn, 2000). Later, the RSPB and WWF joined forces to lobby EU institutions (Pritchard, 2000). Among other things, they provided technical and legal assistance to the senior DG Environment official, Stanley Johnson, who sat down and wrote the first draft of the proposal, and the EP's rapporteur, Hemmo Muntingh, who (it will be recalled) had also been intimately involved in the development of the Birds Directive.

The negotiation of European Union policy

The Commission finally issued a proposal on the conservation of natural habitats and wild fauna and flora in September 1988. Formally speaking, it was supposed to allow the EU as a whole to implement the 1979 Bern Convention (see above), to which individual EU states had already signed up and which had been implemented via national measures. However, the proposal also included a category of endangered semi-natural habitats irrespective of whether they contained threatened species. Sharp (1998) explains that it was because of this particular feature of the proposal – which, in effect, transformed it from being species-driven to encompass habitats – that national environment departments dubbed it the 'habitats directive'. The Commission's bold attempt to deepen integration suddenly raised the prospect of further misfits, forcing national environmental departments to address the adequacy of their national site protection.

Their Lordships criticized Britain's response to the proposal for being 'unenthusiastic' and 'parochial' (Sharp, 1998, p.37). As with the EIA proposal (Chapter 10), they believed that the DoE should have proactively domesticated the EU by uploading British expertise and standards, thereby strengthening Member States with less advanced biodiversity policies. But at a time when many experts and pressure groups were beginning to treat biodiversity as a shared *European* resource, the DoE was unwilling to think in similar terms. Among many DoE officials there still remained the view that:

> our protective systems were well developed and among the best, so why have all the bother of superimposing a set of European

requirements whose practical benefit would largely be to improve the prospects for nature conservation in more backward member states? Against this background the watchword was to keep a low profile and proceed with caution. All this tempered enthusiasm for the fresh wave of provisions heralded by the new Directive.

(Sharp, 1998, pp.37–8)

These attitudes were accentuated by Ridley's deep philosophical antipathy towards all things European. But there were also other factors at work, not least the pressure exerted on the DoE by cognate departments, who were anxious not to repeat the mistakes in relation to bird protection. Politically speaking, the proposal also jarred with the principle of subsidiarity eagerly promoted by the core executive during the early 1990s (Chapter 5) and the voluntary flavour of pre-existing British policy (HOLSCEC, 1989, pp.15–16). However, Sharp (1998, pp.42–3) recalls that British attitudes softened as the negotiations progressed. He attributes the change of heart to the DoE's efforts to make the proposal less mechanistic and more practical, and to the gradual Europeanization of the DoE post-Patten (Chapter 2).

However, Britain reverted to a more defensive stance when, midway through negotiations in the Environment Council in 1991, the ECJ issued its judgment in the Leybucht case (see above). British negotiators were deeply concerned that the judgment, 'which confirmed the worst fears of economic departments and the European lawyers in Whitehall' (Sharp, 1998, p.38), would limit the scope for developing SPAs. At first blush, the ECJ's ruling implied that the development of an SPA could *only* be justified on grounds of human health and safety. No Whitehall department, not even the DoE, was prepared to accept this level of external constraint, since it would effectively render all SPAs *inviolable*. This ruling also made many previously uninvolved Whitehall departments sit up and ask the DoE much more searching questions about what it was agreeing to in Europe.

In an attempt to foreclose future misfits, the DoE set about framing a series of amendments brokered in one of the Cabinet Office's European coordinating committees. These were intended to make the proposal more discretionary and to extend this approach back to the Birds Directive. But the DoE found itself opposed by those who had benefited from the EU's involvement (i.e., policy feedback leading to societal lock-in). The Commission and several environmental groups were particularly dismayed by Britain's attempts to subvert the Leybucht ruling. Nevertheless, the DoE, backed by several other national environment departments, persevered and the initiative was adjudged 'successful' (Sharp, 1998, pp.39, 40). Indeed, the main sticking point during the endgame turned out to be the provision of financial support to Southern Europe states.

Formal compliance

The Habitats Directive was formally adopted in June 1992. It places obligations on Member States to protect plant and animal species and their habitats by merging SPAs with a new class of areas known as Special Areas of Conservation (SACs) to form a pan-European system of protected areas called Natura 2000. British efforts to prevent deeper Europeanization are apparent in Article 2, which requires states to protect habitats whilst taking account of economic, social and cultural requirements, and regional and local characteristics (cf. Article 2 of the Birds Directive). Moreover, Article 6(4) permits environmentally damaging developments to proceed if there are 'imperative reasons of overriding public interest', including those of a social and/or economic nature,[5] and adequate compensatory measures are provided to maintain the overall integrity of Natura 2000. The DoE had hoped to rely on existing town planning and nature conservation legislation to implement the Directive, but in the light of adverse ECJ decisions and the omnipresent threat of future infringement proceedings (such as Case C-335/90), it decided to issue legal regulations instead. The 1994 Conservation (Natural Habitats, etc.) Regulations are an almost direct transposition of the Directive into British law.

The guidelines for identifying SACs are less ambiguous than the relevant parts of the Birds Directive, but implementing them has still proved to be problematic (Ledoux *et al.*, 2000). The DoE sent its first list of 136 SACs to the Commission in June 1995. A second list of 75 terrestrial and 10 marine 'candidate' sites was sent in January 1996. Following strong hints from the Commission that many more were expected in order to implement the spirit of the Directive, the DoE submitted a much longer list. By 1999, no fewer than 340 candidate SACs had been identified but the number of marine sites remained comparatively low, with only 38 SACs and 69 SPAs respectively (HOLSCEC, 1999, pp.9 and 14). Rulings by the ECJ have confirmed that EU states should indeed nominate *all* eligible sites based solely on a scientific assessment and not (as was the case with many national systems of protection) administrative or political convenience. Currently, Britain lies towards the bottom of the unofficial 'league table' of the EU, with 7.4 per cent of its total territory classified for special protection (see Table 8.4). A more recent

Table 8.4   SACs notified to the Commission (as of November 2000)

Member State	Number of sites classified	Total classified area (km²)	% of national territory
Belgium	209	1,105	3.6
Denmark	194	10,259	23.8
Germany	2,196	20,434	5.8
Greece	234	26,522	20.1
Spain	937	90,129	17.9
France	1,028	31,440	5.7
Ireland	317	6,140	8.7
Italy	2,507	49,364	16.4
Luxembourg	38	352	13.6
The Netherlands	76	7,078	17.0
Austria	127	9,144	10.9
Portugal	94	16,502	17.9
Finland	1,381	47,154	13.9
Sweden	2,454	50,908	12.4
UK	386	17,941	7.4
Eur15	12,178	384,472	

Source: http:// www.europa.eu.int/comm/ environment/news/natura/index_en.htm

assessments undertaken by WWF (2001) painted Britain in a more favourable light, but maintained that no state was performing strongly.

Practical implementation and policy impacts

In its rulings on bird protection, the ECJ strongly supported ecological protection. However, the Habitats Directive has helped to swing the balance back in favour of national autonomy. For instance, in 1995 the Commission relied upon the 'overriding public interest' clause of the Directive to sanction the construction of an environmentally damaging road in North Germany, effectively privileging the EU's transport and regional policies over its environmental policy (Nolkaemper, 1997, pp.273, 277). These changes notwithstanding, substantial and largely unforeseen misfits have arisen in Britain. The absence of any pre-existing national controls on the protection of marine areas has proved to be especially troublesome. Throughout the negotiation of the Directive, the DTI and MAFF fiercely resisted any extension of the Directive to marine areas (echoing their attitude to the Birds Directive), even though the Directive makes no such distinction. Lawyers in Whitehall warned that the inter-departmental compromise brokered though the Cabinet Office – an administrative distinction between marine and non-marine areas – was unlikely to hold up in a court of law, but they were not taken seriously enough. These fears were soon borne out when, in 1999, Greenpeace challenged the Government's interpretation in the High Court. The Court ruled that the Directive did in fact apply to the continental shelf and to waters up to the 200-mile fishing limit, rather than 12 miles from British shores. In October 2000, the DoE was forced to issue new implementing regulations, which will impose entirely new environmental constraints on future oil and gas exploration in the North Atlantic (ENDS, 299, pp.54–5). The British Environment Agency is also beginning to appreciate the extent of the legal misfit with existing practices. The problem is that the Directive requires *all* existing discharge consents, water abstraction licences, sea defences and planning consents that affect protected sites to be consistent with EU biodiversity protection requirements. The Environment Agency has had to be granted new legal powers to undertake this massive rewriting operation.

Theoretical reflection

When the Commission began to look for opportunities to expand the biodiversity *acquis*, Britain sought to shelter economically powerful domestic groups (i.e., agriculture, landowners, and businesses) by ensuring that EU policies fitted with pre-existing national rules and underlying philosophies. More specifically, it sought to secure three main objectives: (1) to maintain the pre-existing nature protection regime, which relied on a mixture of national initiatives and international conventions dating back to the 1950s; (2) to preserve the essentially voluntary approach to conservation that was well established in Britain; (3) to bring 'laggard' Member States up to its own 'high' standards of nature conservation.

The evidence presented above suggests that none of these has yet been fully achieved. Taking each in turn, Britain would have had to produce some form of extra bird protection legislation in the early 1980s in order to implement the new international commitments it had undertaken (Haigh, 1989a, p.296). But this would have been very different in form to the Birds and Habitats Directives, which are formal, less discretionary and inherently expansive in their scope. The British government, and in particular the DoE, failed to appreciate that European commitments are different in their stringency and scope to those entered into under international law. Second, although Britain was at a comparatively advanced stage in the development of biodiversity protection policies, they were always subservient to national economic and social requirements. EU policy goes considerably beyond these pre-existing national measures in two important respects: it constrains previously unconstrained economic activities such as agriculture, oil and gas production and road building; and it encompasses species and their habitats as an integrated policy problem. Third, at one time or another, almost every state has been charged with not implementing the two Directives. This implies that Europeanization in the laggard states has not yet pulled them up to the same level as the leaders. In summary, the basic paradigm underpinning British nature conservation policy (voluntarism), the tools used to achieve policy objectives (voluntary agreements and weakly enforced regulations) and the process through which these tools were calibrated (quiet negotiation), have all been extensively transformed by the EU (see Table 11.1).

How well do these longer-term policy outcomes relate to the theoretical predictions made in Chapter 3? Stripped to its core, LI argues that states fight to secure their economic interests in Europe, yet the standards that Britain has eventually been forced to accept in this particular policy area are inimical to the economic interests of the agricultural policy community and those wishing to develop rural areas. They are also incompatible with existing British nature conservation policy and practices. Moreoever, these outcomes arose in spite of the efforts made by British authorities to domesticate EU policy by engaging in partial or selective implementation. The DoE did not stand idly by as the two policies steadily Europeanized British policy, but opposed the Commission in several cases brought before the ECJ. It also tried to bargain with DG Environment for a more minimal interpretation of Britain's legal obligations under the two Directives. Up to a point, Britain succeeded in neutralizing the pro-integration rulings of the ECJ by having the Birds Directive amended. But, overall, national autonomy has been greatly reduced for no obvious economic or political benefit. Recent legal developments have added to Britain's difficulties by extending environmental controls to marine areas. This saga has severely tested the well-established alliance between MAFF and the agricultural policy community and has generated intense disputes between different parts of the British core executive. Finally, it is very difficult to argue that the British government (or even the DoE) used the EU to strengthen itself or domestic policy at the expense of societal groups (i.e., slack cutting) or cognate departments (i.e., home running). Rather, it spurned the opportunity to upload its relatively advanced national policies and philosophies to Europe, and then struggled to cope with the consequences.

Process-based theories, on the other hand, would interpret the unexpectedly profound Europeanization of British biodiversity policy as an unintended outcome of a series of seemingly 'small' decisions. Three aspects of the whole saga stand out as being especially noteworthy. First, process did matter. At first sight, the wild birds proposal seemed unthreatening, not least to Peter Shore. However, the DoE failed to anticipate the long-term consequences of delegating authority to the EU. Before long, Whitehall found itself operating within a steadily expanding framework of rules that constrained its autonomy and limited future choices. The speed at which the biodiversity *acquis* was transmogrified took the whole of Whitehall (including the DoE's Legal Group) by surprise. During the all-important formative moments of EU policy, British negotiators failed to appreciate that protecting birds would inevitably lead to scientific and political demands from large, well resourced but previously marginalized pressure groups (such as the RSPB) for habitats to receive the same level of protection: demands that the Commission was eager to foment and only too pleased to respond to.[6] Over time, the Europeanization triggered by the Directives generated significant societal adaptation (i.e., the politicization of pressure groups and the creation of new political opportunity structures), which helped to lock the new European rules into place.

Second, the preferences of the national actors working within these rules shifted as they became more involved in European rule-making and implementation. The change that occurred within the DoE is especially noteworthy. Economic pressures on the department were comparatively weak and it was left to negotiate the Directive relatively unsupervised by other departments, who naively expected European rules to fit automatically with national ones. However, when European birds legislation was first mooted, the Department was still overwhelmingly national in its activities and expectations and so it spurned the initial, 'golden' opportunity to domesticate European rules from the outset. Consequently, the Department can claim little credit for the environmental benefits brought by the gradual Europeanization of national policy. In fact, its miscalculations (or at least false expectations) about the long-term impacts of European integration raise serious doubts about whether it understood fully the nature of the commitments into which it was entering. In time, though, the DoE has begun to appreciate, almost serendipitously, that it can achieve higher environmental standards using European rules than via national channels of action, which had always been dominated by MAFF and the DTI. However, there is no evidence that the DoE deliberately set out to achieve these outcomes in the 1970s by 'home running' in the EU.

Third, supranational agents played an active part in advancing integration and Europeanization. The Commission, led by political entrepreneurs such as Clinton-Davis and Johnson, was very adept at employing 'subterfuge' to expand the scope and stringency of European biodiversity policy. The Commission's tactics included generating consensus by convening groups of national experts, piggy-backing European rules on international conventions, and sharing technical information with national and European pressure groups which functioned as its 'eyes and ears' at the national level. The expansion of the *acquis* was supported and, on occasions, bolstered by ECJ rulings. By the early 1990s, supranational actors, encouraged and on occasions directly supported by national and international pressure groups, had succeeded in advancing a much more maximal interpretation of EU law than states had originally expected. Today, states are unable fully to control economic and social activities within their own sovereign territories. In Britain, sub-national government authorities (such as English Nature) are formally responsible for identifying European protected sites, and the governing framework of rules within which they operate is no longer national, but *European*.

9

Air Policy: The Air Quality Standards and Integrated Pollution Prevention and Control Directives

For reasons outlined more fully below, the EU's air quality policy took a lot longer to get going than its water pollution policy. Crucially, the Commission only really began to take significant strides in the period *after* Germany's post-1982 'conversion' to air quality improvement (see Chapter 2; and Weale, 1996). Today, however, the air quality *acquis* encompasses well over 50 separate items, including some of the most costly pieces of EU environmental legislation ever adopted, as well as measures addressing car pollution and also climate change. However, this particular chapter examines the origins and adoption of just two items, namely the 1980 Directive on Smoke and Sulphur Dioxide (80/779/EC) and the 1996 Directive on Integrated Pollution Prevention and Control (96/61/EC), and examines their unfolding impact in Britain. Strictly speaking, IPPC cuts across water, land and air but, as will become clear, its origins lie in a synthesis of British and EU air pollution policy. Spanning, as they do, two quite different phases of European regulatory activity (i.e., the 1970s–1980s and the 1980s–1990s; see also Chapter 2), they also reveal how far the DoE's relationship with the EU has evolved in the last 30 years.

Historical background

The early development of EU air pollution policy was hindered by three factors: the protracted paradigm dispute in the water sector (which absorbed vast amounts of the Commission's time and soured political relations: see Chapters 2 and 7); the 1973–4 oil crisis (which made states extremely reluctant to intervene in energy matters); and conflicting national approaches to air pollution control. In the early 1980s, concerted German efforts to upload national air pollution policy provided EU policy with the fillip needed to shake off these constraints. Prior to 1982, the Commission had been forced to concentrate upon areas with a strong single market (e.g., the 1975 Directive on the sulphur content of fuel oils) or, as in the case of some water and air quality standards, the human health dimension. Consequently, European integration proceeded relatively slowly. For instance, EU air quality standards (AQSs) were set for sulphur dioxide in 1980 after four years of negotiation, and for lead (82/884) and NiO_2 (85/203) in 1982 and 1985 respectively.

The DoE adopted a very reactive approach to the Smoke and Sulphur Dioxide Directive, even though the Commission's initial proposal drew heavily on the work of a renowned British epidemiologist, Professor Lawther. Britain had a long history of air pollution control dating back to the mid-nineteenth century, and the DoE's Air, Noise and Waste Division felt it could easily adapt to any regulatory demands made by the EU, which was only just starting down the road of environmental management. To varying extents, this attitude of mind was also present in the water (Chapter 7) and planning (Chapter 10) divisions. But for reasons which will become clearer, the Europeanization generated by the Smoke and Sulphur Dioxide Directive has been significantly less than that produced by EU legislation in cognate fields such as water and biodiversity protection. Put simply, an unambitious European policy and a very muted response from national pressure groups combined to produce a fairly close fit between EU and British practices, limiting the depth of Europeanization (see Table 11.1).

The story of how the DoE handled the IPPC Directive is different but no less intriguing. It originally stemmed from a British attempt to upload a 'good' idea to the EU. However, once inside the EU system, the EPG gradually lost control of the situation as its notion of integration was at first supplemented and then eventually transformed by similarly 'good' ideas uploaded by other states. The original British variant of IPPC, IPC (Integrated Pollution Control), which the DoE uploaded, addressed the emission of pollutants from certain highly polluting factories, whereas European IPPC looks at a much wider range of factors including material inputs, noise pollution, energy efficiency and waste minimization (Emmott and Haigh, 1996). A British policy exported to the EU was thus transformed by the EU in ways that surprised the DoE.

Pre-existing British Policy

Britain has a long and distinguished history of regulating air pollution from stationary sources dating back to the 1863 Alkali Act. The IAPI was at the heart of a professional policy community of pollution control experts comprising chemical engineers, factory owners, scientists and senior DoE officials. It tackled the most highly polluting emissions by collaborating closely with owners to achieve a phased programme of reductions following the principle of best practicable means (BPM; i.e., that pollution should only be reduced where and when it was economically and technically practicable). The responsibility for addressing less polluting emissions such as grit, smoke and dust continues to lie with local environmental health officers, who use various statutory (the Clean Air Acts) and non-statutory tools to improve urban pollution. Under the Clean Air Acts, local authorities had the power to issue 'smoke control orders' prohibiting the emission of smoke from buildings. The whole system was highly localized insofar as the SoSE had no general power either to set emission standards for the most polluting substances or to set local AQSs for the lesser polluting ones. These diverse elements were underpinned by a policy paradigm which suggested that regulators should concentrate upon minimizing the known effects of pollutants after they had entered the environment, rather than eliminating all emissions regardless of their impact. In effect, scientific proof of actual damage done to the environment (but especially human health) was deemed to be the *sine qua non* of effective pollution control. This 'contextual' approach (which also underpinned British water policy) to dealing with pollution, was staunchly defended by the scientific and political establishment in Britain, including the RCEP and the highly respected House of Lords Select Committee on the European Communities.

The Air Quality Standards Directive

The origins of EU policy

Sulphur dioxide was listed as a 'first category' pollutant in the First Environmental Action Programme. In 1973, the Commission's social policy Directorate (then DG V[1]) responded to this obligation by establishing a working group of national experts to develop a proposal to limit atmospheric concentrations. The DoE sent Professor Lawther to represent Britain in the Commission's working group. On paper, Lawther was a potentially powerful advocate of national policy, because he had recently drafted WHO guidelines on sulphur dioxide which the Commission was expected to take as its point of departure. From the start, the risk of serious misfits, therefore, appeared low. The group was apparently united in its concern about safeguarding human health (especially in urban areas) rather than protecting the environment for its own sake (cf. Germany post-1982). This fitted the British agenda perfectly. After much discussion, the Commission decided to push for a two-stage approach, with one Directive tackling the sulphur content of heating oil (i.e., one of the sources of pollution) and another setting AQSs for ambient concentrations of sulphur dioxide in urban areas. These two elements were supposed to function together as an integrated package, with the former providing one of the means of achieving the latter. However, the sulphur oil proposal ran into a barrage of opposition from Member States who were in the midst of an oil crisis, and it was eventually ditched, leaving the AQS Directive to function on its own.

Quite by chance, around this time, the RCEP was undertaking a major review of British air pollution policy. It held meetings with DG V officials and took advice from Professor Lawther. In its report, which was published in January 1976, the RCEP endorsed the basic philosophy of BPM but recommended that it be extended to cover all emissions to land and water as well as air (Jordan, 1993), to prevent waste from moving from one medium to another. So was born the British concept of IPC. The RCEP (1976, p.109) also recommended that local authority powers be significantly strengthened, but condemned the Commission for relying too heavily on uniform AQSs. Although clear and seemingly stringent, the RCEP claimed that such standards were unenforceable, because at any one time there may be many hundreds or even thousands of separate sources of pollution. Identifying which one (or ones) should bear the brunt of the clean-up needed to attain a local AQS would, their Lordships claimed, be an extremely difficult, if not impossible, task. There was also the problem of where to site monitoring equipment given the inherent variability of ambient air quality (1976, p.171). The RCEP therefore recommended the DoE to domesticate the EU by advocating an approach that fitted prevailing British practices. In this approach, AQSs would be expressed as flexible guidelines that could be adjusted upwards and downwards along a series of bands to reflect local conditions and circumstances (1976, pp.169–78), fitting the philosophy of BPM. It concluded with the prophetic observation that DG V was:

> generally progressing too quickly without allowing time for basic research or even for assimilation of the results of existing research

and discussion with experts . . . We do not think that . . . imposing rigid, statutory limits is either wise or practicable. Such limits would be unenforceable in practice and would bring the law into disrepute.

(1976, pp.182–3)

The negotiation of European Union policy

Undaunted by these criticisms, DG V issued a proposal for statutory AQSs for sulphur dioxide and smoke in February 1976. Although the proposal conflicted with the flexible policy paradigm underpinning British practices, the DoE believed that it would not be unduly difficult to implement given Britain's 'proud' history of controlling air pollution (Hajer, 1995). No attempt was made by the DoE to upload the RCEP's movable band approach to the EU. Instead, the SoSE, Dennis Howell, simply assured British MPs that Britain 'led the world' with the 1956 Clean Air Acts (HC Debates, Vol. 932, 18 May 1977, col. 602) and would continue to do so after the EU's involvement. Because the Commission's proposal drew upon the scientific work of Lawther, the then Head of EPG's Air, Noise and Waste Division, James Batho, said that the DoE 'do[es] not really wish to oppose [it] fundamentally' and agreed with it 'in principle' (HOLSCEC, 1976b, p.28). Batho reserved his harshest criticism for the (ultimately) doomed sulphur in oil proposal, which the DoE appears to have regarded as the more problematic (and hence threatening) of the two linked proposals (1976b, p.37).

Their Lordships agreed with the RCEP's warning that the Commission's approach was fundamentally 'misguided' insofar as ambient concentrations could not easily be traced back through the environment to particular sources. They also criticized DG V for inserting Lawther's work into a framework that 'misfitted' with British practice, and suggested that between 5 and 10 per cent of urban areas would not comply with the proposal. Most of these lay in areas of Northern England where the domestic consumption of coal was still high. Lawther admitted that 'after all the [scientific] work we put in . . . they [DG V] really have made a mess of it' (in: HOLSCEC, 1976b, para. 1). But in some respects, the DoE's failure to domesticate the EU was not that significant because:

pollution has decreased to such an extent that we are among what we call in the trade, background noise . . . The [proposed] criteria . . . are in our favour. We are almost always underneath the[m] except . . . on days when nobody can stop pollution being high because pollution is dependent . . . on the vagaries of the weather.

(Lawther, in HOLSCEC, 1976b, para. 1)

For reasons that are still unclear, their Lordships' criticisms were never actually debated in the House, and so the DoE was never called upon publicly to justify its acceptance of the proposal. Consequently, a paradigmatically different but unthreatening (as least to British practices) EU policy was allowed to sail through the British legislative process almost entirely undebated.

However, negotiations on this dossier in the Council working groups dragged on for over four years. The DoE was particularly concerned about the compliance deadlines, which it managed to have put back (HC Debates, Vol. 932, 18 May 1977, col. 603), and some of the limit values, which it managed to weaken. In July 1980, the Environment Council finally adopted the proposal. Knill (1997, p.50) suggests that bargaining in the Council successfully whittled down the Commission's original proposal to achieve a goodness of fit with national practices. Shortly after, a DoE official acknowledged that the Directive marked 'a significant turning point' in the British air pollution policy paradigm, but maintained that Europeanization would be limited because 'there is no intention to seek to control emissions on a short term basis to prevent breaches. The necessary action must inevitably be long term changes in controls or emission standards' (ENDS, 61, p.3).

Formal compliance

The Directive sets AQSs to protect the environment and safeguard human health. Article 3 requires states to take 'appropriate measures' to ensure that limit values were met by 1983, with full compliance by April 1993. The limit values, which must not be exceeded for certain set periods of time, relate to the two main pollutants, SO_2 and smoke. Article 2 also implores states to attain stricter but non-binding guide values by designating special improvement zones, but there are no compliance deadlines. Under Article 4, states can set tougher standards in environmental improvement zones. The original deadline for submitting these plans to the Commission was April 1983. Interestingly, unlike most other environmental Directives, the enforcement regime is extremely anomalous because states need only report sites which breach limit values to the Commission (which must then assume that the remainder are compliant: CoM (95) 372). Finally, the Directive requires states to locate their sampling equipment in the most sensitive (i.e., polluted) areas, although it has never had an independent means of checking this.

Initially, the DoE sought to limit Europeanization by transposing the Directive in the simplest way possible (in March 1981 it issued an administrative Circular: DoE, 1981). It also recommended a very *de minimis* approach to implementation by advising local authorities against establishing either type of improvement zone where higher standards applied. Two years later, it informed the Commission that 29 district councils were non-compliant and promised full implementation by 1993. Most of these sites were, as the RCEP rightly predicted, located in the coal communities of West and South Yorkshire, where free coal had traditionally been distributed to miners. Domestic smoke controls had always been very weakly applied in these areas so as not to render the coal unusable.

In 1985, the Commission issued its first comparative implementation report, which revealed that the DoE had failed to submit adequate plans to remedy the 29 cases of non-compliance (CoM (85) 368 final). By the late 1980s, the number of non-compliant areas had dropped to 22. At one point, the Commission was rumoured to be on the verge of initiating infringement proceedings after receiving a letter from a resident living in the Southwich district of Sunderland (*The Independent*, 30 May 1989), but apparently the circumstances were not serious enough and it decided not to push the point with the DoE. The Commission also pointed out that the limit values had to be enshrined in national legislation and commenced preliminary legal proceedings. Here, the Commission was on firmer legal ground and the DoE was eventually forced (DoE, 1986a) to introduce new primary legislation to establish national AQSs, although these powers were not actually adopted until 1990.

Practical implementation and policy impacts

A second implementation report issued by the Commission in 1995 revealed large variations in the way states monitored pollutants and designated improvement zones (CoM (95) 372, 15). Some had not designated a single improvement zone under Article 4 and the guideline values appeared to be a dead letter in most states. Meanwhile, some states appeared to follow the spirit of the law and monitor in the most polluted zones, whereas others have only monitored the cleaner areas.

In spite of the DoE's attempts to reduce the misfit with national practices, the Directive has altered the tools and style of British air pollution control policy. First, it forced the DoE to become much more involved in matters that had previously been devolved to local officials. Second, the

Directive has undoubtedly accelerated the implementation of existing national laws governing air pollution, sharpening tensions between national and sub-national government. Originally, the DoE seemed confident that it could meet EU standards without intervening in local matters (P. Evans, 1980, p.3). However, legal pressure from the Commission forced it to complete the air improvement programme initiated in 1956. In that sense, the Directive provided an impetus and a strategic framework for the completion of national policy programmes.

Third, the Directive introduced AQSs into British policy, breaking a tradition of informal control dating back over a century. These standards were subsequently enshrined in the 1990 Environmental Protection Act, following pressure from the Commission. To that extent, British law and policy have been Europeanized in ways that were not originally foreseen or desired by the DoE. In the 1990s, the Commission proposed a new air framework Directive to consolidate and remedy many of the failings (see CoM (95) 372) of the existing AQSs Directives. This was finally adopted in 1996 (96/61/EC) and will generate a host of new 'daughter' Directives covering separate pollutants. The DoE fought for a much looser and more devolved system of control, but was eventually forced by other states into accepting much more intensive monitoring and a longer list of 'priority' pollutants.

Since then, the popularity of AQSs as a tool of environmental management has increased significantly in Britain; so far, in fact, that British politicians have used existing EU AQSs as a vehicle to achieve even stronger environmental policies through domestic legislation. For instance, when he was SoSE, Gummer issued an ambitious National Air Quality Strategy with the aim of progressively improving the quality of urban air. The CBI and the chemical industry criticized it for setting standards that were *more* stringent than those under discussion in the Commission (ENDS, 262, p.33). When it entered power, Labour promised a 'right to clean air' in Britain and produced an even more ambitious national air quality strategy (DETR, 2000b), which went beyond EU requirements.

However, the *practical* impact of the original 1980 Directive has been much less significant than some of the other 'small' decisions analysed in this book. First, air pollution levels were already falling (mainly because of economic and technological improvements) when the EU standards were adopted. Average concentrations of sulphur dioxide have decreased five-fold since 1960 (DETR, 1997, p.161). Gooriah and Williams (1982, p.11) calculated that 925 sites would have failed EU standards in 1962/3, 58 in 1979–80, 11 in 1980–1 and 81 in 1981–2 (the coldest winter last century: see Figure 9.1). The vast majority of these breaches were due to excessive amounts of smoke. Today, there are virtually no breaches (see Table 9.1), although smoke and SO_2 concentrations remain at a high level in some areas (e.g., Northern Ireland) where coal is still used for domestic heating (DETR, 1997, p.175).

Second, Britain tended to site its monitoring equipment in areas of low pollution, on the grounds that the Directive addresses continuous exposures rather than short-run exposures in polluted city streets. The DoE also made maximum use of the implementation period and maintained a very *de minimis* approach to compliance.

Perhaps realizing the drawbacks of AQSs, in the 1980s, the Commission began increasingly to address the sources of air pollution, rather than tackling their after-effects. Although this was consistent with Britain's desire to concentrate on those emissions that actually generated pollution, the Germans advocated much faster, deeper and comprehensive cuts than Britain was prepared to accept. The conflict spiralled into a major international dispute between Britain and the rest of the EU over acid rain, which brings the story neatly to the concept of IPPC.

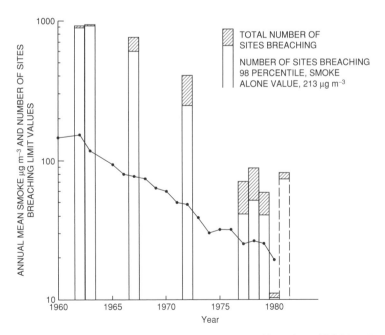

Figure 9.1 UK urban average smoke concentrations and breaches of EC Directive 80/779
Source: Gooriah and Williams (1982), p.18.

Table 9.1   Exceedences* of the EC Directive, 1983–97

Year	No. of sites exceeding any limit	Site names
1983	12	Goldthorpe 1, Grimethorpe 2, Wombwell 2, Whitehaven 2, Askern 6, Doncaster 27, Doncaster 32, Moorends 1, Mansfield Woodhouse 2, Sunderland 8, Castleford 9, Ashington 4
1984	6	Belfast 12, Belfast 17, Belfast 33, Londonderry 8, Newry 3, Newry 4
1985	9	Grimethorpe 2, Barnsley 9, Doncaster 29, Doncaster 32, Moorends 1, Askern 8, Sunderland 8, Belfast 12, Newry 3
1986	5	Langold 1, **Seaham 2**, Hetton-le-hole 3, Belfast 13, Newry 4
1987	9	Crewe 17, Mexborough 19, **Seaham 2**, Sunderland 8, Hetton-le-hole 3, Houghton-le-spring 2, Featherstone 1, Ashington 4, Belfast 39
1988	3	New Ollerton 2, Hetton-le-hole 3, Belfast 39
1989	3	Hetton-le-hole 3, Belfast 42, Newry 3
1990	3	Belfast 11, Belfast 12, Belfast 42
1991	2	**Durham Sherburn 1**, Belfast 42
1992	1	Grimethorpe 3
1993	0	-
1994	0	-
1995	0	-
1996	0	-
1997	0	-

*All sites (except those in bold) are in areas that had a derogation from the EC Directive until 1993. Only those shown in bold constitute statutory exceedances.
Source: DETR (1999).

The Integrated Pollution Prevention and Control Directive

The origins of European Union policy

Three actors helped to place IPPC on the EU's political agenda. It is indubitable that the Commission had harboured a desire to create an EU-wide system of integrated licensing (i.e., one permit covering emissions to air, water and land) well before it issued a proposal on IPPC in 1993. Following Germany's sudden environmental 'conversion' in 1982, the

1984 Air Framework Directive introduced the German notion of Best Available Technology Not Entailing Excessive Costs (BATNEEC) into British policy which, in time, superseded BPM (Jordan, 1993) as a means of reducing acid rain (see above). However, opposition from the British (among others) effectively reduced the Commission-directed process of agreeing 'daughter' Directives for individual pollutants to a snail's pace. The second actor was the DoE. Steps had been taken in Britain throughout the 1980s to establish a national system of IPC based on the RCEP's 1976 plans (see above). IPC was eventually implemented 14 years later by the 1990 Environment Act, which was some three years after the establishment of an integrated permitting agency, HMIP, in 1987. IPC was the EPG's baby, delivered in the teeth of concerted opposition from other Whitehall departments (O'Riordan and Weale, 1989) so, when the DoE began to pursue a more engaged European strategy in the early 1990s, the EPG's high command immediately seized on IPC as just the sort of national innovation that could usefully be uploaded to the EU. According to one senior official in the Air, Noise and Waste Division, uploading 'good' ideas to the EU had an appealing logic:

> If you've got a good idea nationally and if you put it into Europe you can lumber all your competitors with the same moans you're getting from your own industry. This is exactly what we did with [IPC]. Once we passed the 1990 Act, I seconded over to the Commission the same [civil servant] who had done the bill and he drafted the [IPPC] Directive, which oddly enough looks like our IPC system!

The DoE was able to demonstrate its ability to 'think (and act) European' by operating at several different levels of governance simultaneously. It started by working inside the OECD's environment group, which continuously drips ideas into the Commission. In 1991, this group duly issued a recommendation on IPPC (OECD, 1991), where the additional 'P' stood for 'prevention' (reflecting the continental European preference for source-based controls). Thereafter, the OECD established a working group to look at the practical difficulties of implementation (CoM (93) 423), but even at this very early stage it was obvious that EU IPC would not be identical to British IPC.

By also working at these three levels of governance, the third actor, the IEEP, helped the DoE to complete the export process. For example, it organized a conference in Brussels which discussed various approaches to IPC and published a book containing practical examples of national IPC systems (Haigh and Irwin, 1989). It also produced a report for DG

Environment, which recommended integrated permitting and a network of national inspectorates rather than one central European inspectorate (Haigh, 1989b). This report was discussed at a meeting of national experts convened by the Commission in May 1991 to discuss the case for 'integrated permitting' in the EU. Although there were still many administrative hurdles to be overcome – not least the wide variation in national approaches – this meeting demonstrated that the basic concept of IPC had, by now successfully taken root in Brussels (ENDS, 196, p.35). At this point, the EPG's secondee began to insinuate the DoE's ideas into the Commission, first by producing a discussion paper on integrated permitting, which the Commission published in the summer of 1991, and then a working draft of a legislative proposal (CoM (93) 423), which a Commission working group of national experts accepted during the Autumn of that year (ENDS, 201, p.33).

The key problem for the Commission was how to define specific emission standards for different industries without intervening too directly in 'national' affairs. Initially, it wanted to base them on the Best Available Technology (BAT) in Europe, thereby removing the Not Entailing Excessive Costs ('NEEC') element of BATNEEC which was so favoured by the British. BAT implied a more harmonized approach to standard-setting, whereas NEEC was closer to the British concept of BPM insofar as it brought local economic conditions into consideration. However, BAT would have required the Commission to define the 'best available' techniques for different types and classes of industrial operation. Not surprisingly, the DoE was not the only national environment department to feel that this would violate the principle of subsidiarity.

The IPPC proposal emerged slowly out of the interaction *between* these three actors. It was certainly not the Commission's brainchild. In fact:

> When I proposed the Directive . . . There was some nervousness among Commission officials that this was not appropriate for a Directive. They felt it might be regarded as interfering with national administrations since it would not really be setting out 'the results to be achieved' but setting 'the form and methods'.
>
> (Haigh, 1995b, p.4)

However, several states were keen on IP(P)C, so once it was on the EU agenda DG Environment tried to use it as a vehicle for some of its other pet projects: for instance, early drafts included plans for a public inventory of industrial releases and a system of incentive charging, a possible precursor of environmental taxation.

Generally, though, the proposal bore the strong imprint of British thinking, with some phrases and words having been uploaded wholesale from British statutes and policy documents. So, for example, although the DoE had to surrender the 'NEEC' element of BATNEEC to secure the support of greater states, the notion of 'practicability' was retained (ENDS, 201, p.34). The Germans certainly felt the proposal was a British initiative 'incorporating principles from their own long-standing regulations' (Schnutenhaus, 1994, p.323). And yet the basic British-inspired concept of IPC, was quickly transformed in expert discussions as other environment ministries (flanked, no doubt, by their economic minders) sought to reduce the potential misfit with their own systems. In the fifth draft of the Commission's proposal, which was issued in February 1993, IPC had metamorphosed into IPPC, reflecting continental notions of anticipatory pollution control. The inventory and the plans for incentive charging had disappeared, and there was greater scope for states to set their own standards within the broad framework of an EU IPPC system. Nearly $2^1/_2$ years after the creation of the IPC working group, the Commission eventually issued a full proposal for a framework Directive on IPPC (CoM (93) 423).

The negotiation of European Union policy

Their Lordships could not have been that concerned about the Europeanizing effects of the proposal because they decided not to subject it to a detailed scrutiny. Instead, the responsibility passed to a European standing committee in the Commons (HOCESCA). The Explanatory Memorandum submitted by the DoE (1993a, paras 6–7) to the Committee was extremely upbeat about the likely outcome:

> The Government strongly supports the principle of the Directive . . . Insofar as the proposed Directive matches the existing system of IPC in the UK, it should not result in additional costs to industry . . . [T]here is no reason to believe that a move to an integrated approach would necessarily result in any more costly requirements across the board.

However, as negotiations unfolded in the Council working groups, the eventual shape of the final Directive became progressively more difficult to predict. As the dossier mutated, concerns were expressed in Britain about the quality of Parliamentary scrutiny (HOCESCA, 1995, col. 2).[2] A major complicating factor was the arrival of QMV post-Maastricht and the shift towards cooperative decision-making with the EP. Another was the injection of new demands from actors that had been slow to mobilize or had decided to stand by while other actors made the running. Spain, for example, mobilized a coalition of poorer states demanding weaker standards in more robust environmental areas; France made a strong plea for food and drink production to be subject to IPPC; and the Dutch and the Danes also suggested that agriculture and waste management – two highly polluting land uses – should be added to the list. Under concerted pressure from MAFF and the DTI, the DoE opposed the inclusion of agriculture and food production as they had traditionally fallen outside the main scope of national environmental regulation.

By 1995, the proposal was into its eighth draft. In the process, the proposal had metamorphosed from a European variant of British IPC into what one DG Environment official described as an 'extremely moveable feast' (ENDS, 242, p.39). The final negotiation boiled down to a straight fight between a Franco–German model of IPPC (based on uniform emission standards) and a more site-specific form of IPPC (favoured by the British) whereby individual emissions would be tailored to suit local conditions (in effect a European variant of BPM: (see above). When a qualified majority of states supported the inclusion of agriculture, landfill sites and food processing, the DoE set about lobbying MEPs to recover the situation. This required tact and diplomacy because the DoE was anxious to achieve several changes without delaying the adoption of or injuring the very idea that it had championed in the first place. But in a letter to the Vice-Chairman of the EP's Environment Committee, the DoE explained that 'unfortunately, the scope of the draft Directive . . . goes rather wider than the UK's own system . . . [and it] also covers some plant which is appropriately dealt with by neither system' (DoE, 1995, p.1). The letter went on to suggest some 'textual improvements . . . to avoid imposing impracticable burdens on farmers'. The DoE also wrote directly to the sectors concerned, imploring them to lobby MEPs, but to no avail (Skea and Smith, 1998, p.277).

Formal compliance

The Environment Council finally adopted the IPPC proposal in September 1996 after nearly three years of negotiation. The civil servant who headed the British negotiating team said the DoE had 'got what it wanted' from the negotiation (*International Environmental Reporter*, 12 June 1996, p.526). However, in order to secure agreement, the DoE had had to make a number of major concessions with respect to the scope and timing of the new regime, and the stringency of some of the standards. Crucially, the SoSE at the time, John Gummer, was much more willing and able to make these concessions than some of his predecessors, and he did so by horse-trading in the Council: 'you should not expect in every iota and every detail to get what you think is right, because that is not what the Community is about . . . . The [IPPC Directive] isn't perfect, but it is very much better than having separate answers which is totally unacceptable' (Gummer, 2000).

The final Directive is in fact a curious hybrid of British and German approaches. The main requirement is to apply the BAT (German), where 'available' means technically and economically available (British). Crucially, the BAT for a particular operator will, as the DoE demanded, depend upon its technological characteristics, spatial location and economic feasibility (Skea and Smith, 1998).

Practical implementation and policy impacts

In many important respects, the IPPC Directive does not disrupt Britain's basic paradigm underpinning industrial air pollution control (Skea and Smith, 1998, p.280). First, Article 8 explicitly mentions that local conditions must be taken into account in determining BAT, where 'T' is defined as industrial *techniques* (cf. BPM above) rather than the narrow, more Germanic concern with waste reduction *technology*. Second, the DoE also successfully curtailed the Commission's power to propose EU-wide emission limits. Had this clause been adopted, the Environment Council would have been able to adopt emission limits in daughter Directives using QMV (1998, p.276). Because of the changes, the main locus of rule-making in Britain will remain at the sub-national level (i.e., in the site-specific negotiations between operators and British inspectors) rather than in EU technical (e.g., comitology) committees. Third, the DoE achieved much shorter upgrading periods than those sought by poorer states. It also prevented them from relaxing BAT in particular local areas by making the achievement of local AQSs a minimum requirement. Throughout the negotiations, British industrial operators that were already subject to British IPC requirements had complained to the DoE that longer compliance periods and laxer standards would put them at a competitive disadvantage.

These similarities notwithstanding, IPPC is a very different creature to IPC. Indeed, as implementation proceeds, some suspect that both the DoE and the Environment Agency are using IPPC as a vehicle to extend the scope and stringency of IPC. Consider comments made by the Agency's Director of Environmental Protection, David Slater:

It would be easy to adopt a parochial attitude . . . which might justify a minimalist approach to implementation of IPPC. Such a stance, however, would be short-sighted and would overlook the many opportunities afforded by the Directive. In particular [it] can be used as a catalyst for more effective, efficient and holistic control [and] has the potential to become a central part of a suite of policy tools used to pursue sustainable development.

(ENDS, 270, p.21)

Another Agency official maintained that site trials of IPPC had shown that it was much 'more than IPC with whistles and bells, which was the initial attitude of most inspectors and site operators' (ENDS, 292, p.3). These views have alarmed some operators, especially those new to IPC, and brought the Agency into conflict with the DoE, which is anxious not to negate the efficiency gains that IPPC was supposed to deliver. In 1997, the DoE was moved to warn against 'any flamboyant attempt to regulate the entire economy in the interest of sustainable development' (ENDS, 273, p.23). But the Agency is not working entirely alone because, in 2001, several sectors *volunteered* for IPPC after the DoE decided to use the IPPC regime to determine which industries would be eligible for a big rebate on a new national climate change levy (tax).

It is already glaringly obvious that the implementation of IPPC will be considerably more complex and disruptive than first expected. The DoE missed the October 1999 transposition deadline by many months because of the difficulty of fitting together two relatively new and very complex regulatory frameworks: one homespun, the other European (ENDS, 297, p.46). Early indications suggest that IPPC will indeed be a 'meaty challenge even for sectors familiar with IPC' (ENDS, 297, p.3; 298, p.18). In 2001, the Commission initiated infringement proceedings against Finland, Spain, Greece and the UK for delays in transposing the Directive. The DoE, which is unlikely to remedy the breach in the time available, is bracing itself for a critical ruling from the ECJ. It is deeply ironic that the DoE now finds itself alongside two, supposedly 'laggard' states in the race to implement a Directive that it did so much to formulate. So, what practical lessons should the DoE draw from this particular case of Europeanization? Neil Summerton (1999) believes that it: 'demonstrate[s] the danger of exporting a national measure to Europe because what comes out at the other end may be a strange, vestigial creature . . . It perhaps . . . illustrate[s] the importance of being eternally careful in Europe.'

Theoretical reflection

The air pollution sector exhibits many of the same symptoms of Europeanization as water pollution and biodiversity. Over the course of the last 30 years, Britain has had to alter the content, style and underlying philosophy of national air pollution policy to reflect European policy frameworks and ideas. The overall impact of EU policy has not yet been anywhere near as great as in the water sector, but it has been significant nonetheless (see Table 11.1). If we take the AQSs Directive first, the EU has undoubtedly disrupted national practices. There were no AQSs in Britain prior to the EU's involvement so; in that respect, the Directive has had a major and lasting impact on the paradigm of British policy and the tools used to deliver policy outcomes. Among other things, it has made British policy much more explicit, putting legislatively prescribed standards into areas of policy where guidelines and 'rules of thumb' were the convention, and given fresh impetus to domestic initiatives that had stalled. Overall, though, the 'misfit' between the trajectory of national and EU policy was never that great because the standards adopted were weak, so in comparison to water and biodiversity (Chapter 8) the overall extent of Europeanization has been relatively limited. Importantly, the Directive was relatively unambitious with regards to the environmental outcomes sought, so national pressure groups and the Commission never had much to get their teeth into. Thus, there have been far fewer compliance problems than in cognate sectors and the issue never really became that politicized, at least in Britain.

The homespun IPC system has certainly been Europeanized by IPPC, but implementation has not advanced far enough to make any definitive judgements about the overall breadth and depth of Europeanization. However, early indications suggest that the EU effect could be substantially larger than most informed observers originally expected. Crucially, a misfit has arisen because the two systems were designed to achieve different things. IPC was born out of the British tradition of 'reasonableness' and sought to reduce environmental harm in the localities where it manifested itself, whereas IPPC adopts the more continental European (i.e., preventative) approach of fitting the BAT-NEEC to reduce emissions at source, regardless of whether or not they create harm. IPPC also covers many more substances than the 23 listed under IPC, and it addresses very broad types of industrial activity whereas IPC centred on particular processes and substances. IPPC will therefore push controls into areas such as agriculture and food production that were hitherto untouched by environmental regulation. For instance, IPPC provides the DoE with much stronger powers to control ammonia, methane (a potent greenhouse gas) and other air pollutants from the agricultural sector. During cross-Whitehall negotiations, MAFF resisted these and other incursions, but lost (ENDS, 245, p.39). Finally, IPPC pushes source-based environmental controls into the realms of waste and water policy where the EU has previously had little or no presence. For example, until now, site remediation has been imperfectly governed by national land-use planning law. In all these different respects, IPPC has increased national environmental standards and centralized powers that were traditionally exercised more locally while leaving the paradigm and tools of British policy relatively untouched (see Table 11.1). IPPC has also disrupted the closed and secretive style of British pollution control by increasing public access to information.

The main purpose of this book is not however to measure the Europeanization of policy *per se*, but to understand the extent to which it was desired and/or expected by various parts of the British Government, particularly the DoE. Although not a theory of 'small' decisions, LI fits the DoE's handling of the AQSs Directive quite well. The British government succeeded in reducing the potential misfit with national practices by eroding EU policy at the adoption stage. Economic pressures on it were relatively diffuse (compared to the sulphur oil proposal which the oil companies, among others, fiercely opposed), which allowed it to step in ('slack cutting') and surrender aspects of British policy (e.g., the paradigmatic opposition to AQSs) without accepting costly and disruptive EU standards. Shorn of its more Europeanizing elements, EU policy created few compliance problems, excited little political interest and cost relatively little to implement because it worked with the grain of British policy (i.e., the misfit was limited). Supranational entrepreneurship was relatively weak at the agenda-setting and policy-implementation stages, and national pressure groups never showed much interest when possible infringements were detected.

There is, of course, another equally credible interpretation of the same events, which is that these outcomes were due mainly to efforts made by non-British actors. In the late 1970s there was little appetite for air pollution controls in the Environment Council, and the Commission struggled to deepen integration. On this view, the Commission's passivity, coupled with strong state opposition, produced a weak policy that did not misfit much with national practices.

The story of IPC–IPPC is harder to explain in purely intergovernmental terms. It is true that the DoE set out to domesticate EU policy by uploading a national 'success story'. It is also true that it fought to maximize Britain's economic interests. However, ever since it has struggled to mediate Europeanization and integration, both of which have metamorphosed in unexpected and undesired ways. Throughout, EU institutions such as the Commission were not simply instruments of state power. They had their own separate conception of what European integration should mean, enjoyed significant autonomy from the British state and had an independent causal influence on the process of Europeanization.

Interestingly, the Commission's main justification for establishing a European IPPC regime was *environmental*, and not the internal market. That is, it argued that countries which already had integrated permitting systems (namely Denmark, Germany, France, Britain and Luxembourg) or were about to adopt them (i.e., Belgium, Ireland, the Netherlands and Portugal) should not be held back by disintegrated (i.e., single-media) EU legislation (CoM (93) 423, 2). In other words, it used the deficiencies of existing EU legislation in this sector (i.e., single media pollution control) to justify a further deepening of integration. This time, the Commission's incremental approach ('subterfuge' in Héritier's (1999) terms) suited British interests. In the past, Britain would almost certainly have resisted this kind of expansive logic: the chapters on land-use planning and water pollution (10 and 7 respectively) provide ample evidence of this. This tells us two important things about the Department. First, since the 1980s it had become much more environmental: interestingly, the then environment Minister, James Clappison (HOCESCA, 1995, pp.2–3), made it quite clear that Britain wanted IPPC on economic *and* environmental grounds. Second, the DoE had become much better at playing the European game: IPPC reveals how, post-Patten, the EPG used the EU to extend its *environmental* influence on to a much wider (i.e., European) plain. However, in Ken Collins's opinion, the IPPC saga also powerfully reveals how difficult it is for any actor – however 'European' minded – to predict what will eventually emerge from contemporary EU policy processes:

> You might pop something into the system, but the system has other connections and other people are putting things in as well along other tributary pipes. So what comes out of the end is not quite the pure thing you put in in the first place. It will *always* be a negotiated compromise.
>
> (Collins, 2000)

Therefore, when viewed as an unfolding process, it is abundantly clear that process and history did 'matter' in this sub-field of EU environmental policy, and there were significant policy feedbacks. IPC may have been the precursor of IPPC, but it was originally adopted by the DoE to green Britain and rid itself of the 'Dirty Man of Europe' label. In other words, the DoE's interests were shaped substantially by the integration-Europeanization triggered by previous policies. Moreover, IPC was a homespun idea, but it was substantially affected by German thinking, most notably the 1984 air framework Directive and associated acid rain policies. In these and many other respects, the DoE's interests and its strategies were shaped by the environmental *acquis* that had gradually developed around it.

To conclude, although neither Directive has Europeanized Britain to the same extent as some water and biodiversity policies, it is very hard to see how either has strengthened 'the state'. It has almost certainly strengthened the DoE's ability to impose environmental controls on industrial sectors that used their sponsoring departments (MAFF and DTI) successfully to resist environmental controls. The DoE may even have used the IPPC Directive to make a 'home run' against these other departments, though Gummer emphatically denies this (see above). What is clear, however, is that 'the state' as a whole struggled to mediate integration and Europeanization; at critical stages neither process was as predictable (or as controllable) as it had originally expected. Finally, in relation to IPPC, there are clear signs that sub-national agencies do not (cf. LI) simply transpose EU legislation in the automatic and direct manner assumed by state-centric theories. The Environment Agency regards IPPC as an important vehicle to extend and strengthen the national IPC system (i.e., Europeanization), thereby giving added force to EU rules.

10
Land-Use Planning Policy: The Environmental Impact Assessment Directives

If there is one single Directive that exemplifies the DoE's changing relationship with the EU, it is that relating to the environmental impact assessment (EIA) of major development projects. Today, the EIA Directive is one of the most well known of *all* the EU's many environmental policies but, back in the 1970s when the Commission first proposed the idea, Britain vehemently opposed any EU involvement, even though it promised to do little more than formalize many well established features of British land-use planning. When this failed to deter the Commission, Britain pursued a policy of 'pre-adoption emasculation' and then 'post-implementation deflection' to remove any misfits (Wathern, 1989, p.36). This chapter explores the reasons for the DoE's visceral opposition to any EU involvement in an area of environmental policy where Britain so obviously excelled. It also explores the long-term impacts of European integration, as manifest in the Europeanization of domestic land-use planning policy, and the DoE reaction to the new (2001) Directive on Strategic Environmental Impact Assessment (SEIA).

Historical background

The DoE's opposition to EU-level EIA is perplexing given that Britain practically invented the theory and practice of modern land-use planning in the post-war period. Indeed, when the Commission first proposed the idea, EIA was already a fact of life in many major development projects. As a European frontrunner in this area of policy-making, the DoE could have uploaded its experience to the EU, benefiting Member States with 'relatively weak' planning systems (Williams, 1986, p.105).

However, the DoE chose instead to block the Commission rather than sell it a British model of land-use policy. But once it was in place, the 1985 Directive began to generate unintended consequences. No state was more surprised than Britain, which thought it had successfully 'domesticated' EU policy by eroding the Directive to 'fit' the pattern of British practices. The transmogrification of the Directive owes much to the energy of domestic environmental groups, which exploited the new points of leverage (i.e., policy feedback) created by the Directive. According to the former Head of the CPRE, Fiona Reynolds:

> The groups . . . were delighted that the Directive's practical effect would be wider than the UK Government had assumed, and pressed hard for implementation and enforcement to the letter . . . This is partly because EIA is a new process, providing many new 'hooks' on which compliance can be judged, and partly because its timing and importance opened many campaigners' eyes generally to the opportunities presented by lobbying in Europe.
>
> (Reynolds, 1998, p.241)

These campaigns helped to lock the Directive into the domestic political system. The interest the Directive generated in formal systems of environmental assessment (i.e., policy feedback), has also been an important factor in the EU's decision to investigate the utility of an even more comprehensive process of assessment called SEIA. This moves beyond looking at the environmental impact of discrete development projects (e.g., a road or power station) to encompass the overarching policies and programmes of action of which they are part. The EU eventually adopted a Directive on SEIA in 2001 (EC/2001/42), over 20 years after DG Environment first proposed the idea. Interestingly, the DoE had obviously learnt from the EIA saga because it adopted a much more proactive approach to the negotiation of SEIA. By shaping European thinking *before* it became enshrined in legislation, the DoE appears to have foreclosed a number of misfits, although the evidence presented in some of the previous chapters suggests that there may be hidden surprises lurking in the implementation phase.

Pre-existing British policy

EIA is a tool for evaluating the impacts of a proposed development project. Ideally, the systematic collection of information should enable the most environmentally damaging impacts to be alleviated or mitigated entirely. EIA was first introduced by the United States Government in 1969. The success of this system inspired many European states, including Britain, to experiment with similar measures. So, by the time the EU adopted a Directive in 1985, some EU states had had extensive experience with EIA (West Germany, France and the Netherlands, since 1975, 1976 and 1981 respectively), but for the rest, it was still a novelty (Haigh, 1983, p.592; Wood, 1995).

Annex I of the Directive lists nine types of project that must receive an EIA, though exemptions in exceptional circumstances can be made. Annex II includes 13 categories of development covering 80 separate project types which require an EIA where states 'consider that their characteristics so require'. Article 4(2) requires Member States either to specify *a priori* certain types of projects that will fall under Annex II, or establish the criteria and thresholds to determine which apply. In Britain, the main criterion is that the project may generate 'significant' environmental impacts. The Directive may sound extremely intrusive, but it does not actually prevent states from accepting certain forms of development or force them to justify their decisions in public. Rather, it encourages developers to think about (and, where possible, mitigate) the environmental impacts of their activities, and consult affected parties as well as public authorities in the process of decision-making.

Britain developed the world's first comprehensive system of land-use planning in 1948. This bore all the hallmarks of British environmental policy, namely local discretion, pragmatism and case-by-case problem-solving (see Chapter 1). The feeling within the British planning profession prior to the EU's involvement was that the British system was 'mature, comprehensive, workable and, therefore, worthy of emulation as a model for others' (Williams, 1991, p.334). When asked to comment on the Commission's plans for an EU-wide EIA system, one senior county council planner remarked:

> I am convinced we are already on the right lines . . . I do not think it adds to the job that we are doing. I am not convinced that the British people will welcome a directive if it is only to put the affairs of Europe right. Europe must learn from us. They copied us in order to have a parliament like ours so perhaps they had better adopt out planning system . . . [W]e do not want a directive at all.
>
> (HOLSCEC, 1981, col. 248)

Britain did indeed have a mature land-use planning system, but it placed no legal obligation on either developers or planning authorities systematically to consider the wider environmental implications of a particular project (Purdue, 1997, p.244). An EIA was always theoretically possible under pre-existing British law but it was not mandatory, and so in practice very few EIAs were ever undertaken. If a local planning authority insisted on an environmental assessment, there were no national procedures to follow. In fact, the 'case-by-case' paradigm underpinning British policy counselled against the slavish pursuit of common standards.

However, there was interest in EIA long before the European Commission's intervention. For instance, the oil industry in Scotland began to experiment with EIA in the early 1970s to obviate delays in securing planning permission, sometimes with funding from the Scottish Office (Petts and Hills, 1982; Glasson, Therivel and Chadwick, 1994, p.36). Throughout the 1970s, the DoE's Planning Group quietly encouraged developers and planning authorities to innovate for themselves, without formally endorsing EIA or becoming actively involved itself. In 1973, the DoE and the Scottish Office funded a team at Aberdeen University to prepare what eventually became an EIA 'manual', copies of which were posted free of charge to each and every local planning authority in Britain (Wood, 1995, p.45). Local level innovation continued apace in the 1970s; by 1982, over 200 'EIA'-like assessments had been prepared on an *ad hoc* basis (Glasson, Therivel and Chadwick, 1994, p.38).

In 1976, the DoE went so far as to commission a former Under-Secretary to investigate the scope for introducing EIA in the UK (Glasson, Therivel and Chadwick, 1994, p.37). But something must have happened to reduce the DoE's enthusiasm because it delayed the publication of his report, which recommended formal integration of EIA into the planning system, by a year and then distributed it to a much more select audience (Wood and McDonic, 1988, p.13). Under Labour (1974–9), enthusiasm for EIA continued to wane as the economy slipped into a deep recession. Peter Shore believed that EIA could be a potentially powerful tool, but felt it should only be applied to the largest, most controversial projects (DoE, 1978c). He also rejected pleas from British environmentalists for greater EU involvement, warning of: 'the difficulty of translating essentially British ways of doing things into Community wide measures without extending the jurisdiction of the Community into areas which . . . are . . . essentially inappropriate for Community action' (DoE, 1978b). Under the Conservatives, enthusiasm continued to decline. In 1981, the DoE republished the Aberdeen study with a new preface, which warned that '[EIA] should be used selectively to fit the circumstances of the proposed development and with due economy' (Wood and McDonic, 1988, p.13). In practice, many of the criticisms of EIA were always overstated. Experience of applying EIA in the energy sector had shown that often it did little more than attenuate the environmental impacts that the project was pre-committed to generating. More often than not, EIAs were undertaken far too late in the project cycle to address key concerns such as the underlying need for development, possible alternative sites and the cumulative impact of many projects in a bigger programme. SEIA tries to address these problems by ensuring that impact assessments are undertaken at earlier stages in the planning chain (i.e., it tackles the driving forces of environmental damage before they become enshrined in policy and legislation). Like EIA, it also includes formal procedures to encourage consultation and public participation (Therivel *et al.*, 1992, pp.19–20).

The Environmental Impact Assessment Directive

The origins of European Union policy

The Commission desperately wanted EIA to act as the cornerstone of its long-term plan for a more preventative, integrated and environmentally ambitious European environmental policy. The First Environmental Action Programme had not explicitly mentioned EIA or land-use planning, so the Commission knew it had to tread carefully. It began preparing the ground in the mid-1970s by commissioning a number of expert reports. The first was prepared by two British academics, Norman Lee and Chris Wood (Lee and Wood, 1978). In 1975, the EEB tried to capture the unfolding (though mainly nationally-focused) debate on EIA by holding an information-sharing seminar in Brussels (Sheate, 1997a, p.270). On the basis of these discussions, the Commission began to draft a formal proposal in 1977. Straight away, it ran into fierce opposition from states who were understandably reluctant to allow the Commission to intervene in their land-use planning decisions or Europeanize their national EIA systems. This reluctance is reflected in the wording of the Commission proposal (CoM (80) 313), which reportedly went through 23 separate drafts prior to publication in 1980 (Demmke and Unfried, 2001, p.125).

During the drafting stages, important changes were made to accommodate the DoE's objections. For example, Lee and Wood had originally proposed that EIA be extended back up the decision-making process to include broader plans and programmes (i.e., SEIA), but this was quickly dropped (Sheate, 1997a, p.271), as was any reference to agricultural activities in Annex I. However, the final draft still contained a number of elements which the British strongly opposed, namely a long list of Annex I projects where EIA was mandatory, provisions allowing the Commission to coordinate the Annex II thresholds, and a requirement to consider alternative project sites (Wood, 1995, p.32).

The Commission must have known that the political stakes were high because unconventionally, a draft of the proposal was deliberately leaked to the press (von Moltke and Haigh, 1981, p.25). The Union of Industrial and Employers' Confederations (UNICE) declared that it was 'of an importance and scope which no other [EU] initiative on the environment has had so far' (ENDS, 14, pp.7–9). That most influential of British commentators, Lord Ashby of the HOLSCEC and the RCEP, predicted that it would become 'the single most important Directive to come from the [Commission]' thus far (HL Debates, Vol. 419, col. 1320).

The negotiation of EU policy

The Commission had made an extremely bold attempt to upgrade the common interest, so the June 1980 Environment Council's emphatic rejection of its proposal came as no surprise. The British emerged as the most vocal, but by no means the only, opponents. The DoE felt it added unnecessary new administrative burdens at a time when Britain was striving to relax the constraints on economic development. The Planning Group (which coordinated the DoE's response) felt that the 1948 planning system, together with voluntary initiatives by developers, provided a sufficient legal mechanism to allow the largest development projects to receive an informal EIA. It set its face firmly against a formal (i.e., mandatory) system of EIA, encompassing large numbers of different project types. In January 1980, the Planning Group (DoE, 1980a) issued a consultation paper which said that: 'while not opposed to the use of EIA methods in the appropriate circumstances [Britain] is against superimposing an inflexible requirement for compulsory [EIA] on our planning system. On that basis, UK representatives have criticised the Commission's earlier proposals as being indiscriminate, rigid and unsuited to our planning system.'

In private, everything possible was done by the Group to kill the proposal outright. Although shaped strongly by the Planning Group, the British governmental position reflected an interesting amalgam of different departmental viewpoints. MAFF and Energy, backed enthusiastically by the CBI, adopted the most sceptical stance. MAFF was implacably opposed to any extension of statutory control over agricultural operations. It had secured a blanket exemption for all agriculture activities during the drafting of land-use planning legislation in the 1940s (Chapter 8), and wanted that maintained at all cost. During the Parliamentary scrutiny, a member of the British agriculture policy community (Chapter 8), the CLA, even pressed for all forestry and agricultural practices to be formally exempt from EIA (Williams, 1986, p.102). Meanwhile the Department of Energy was concerned that it might delay the construction of new nuclear power stations, which were a key plank of the Thatcher Government's pro-nuclear energy policy. However, other departments, such as transport, defence and Scottish affairs, had had positive experiences with EIA in the past and were at least prepared to be more accommodating.[1]

However, since the lead department (the DoE) was unconvinced, the common position adopted across Whitehall was fairly negative (i.e., to accept EIA in principle but to oppose the Commission's approach to regulating it). According to the DoE's chief negotiator: 'The general DoE view was that the then draft proposals, if handled with care and resolution, were tameable so that the end result would involve no significant additional burdens on the planning system' (Critchley, 1999). In his oral evidence to the HOLSCEC, he said that: '[EIA] is implicit in the whole of our . . . planning system . . . [I doubt] . . . whether one can legislate for it in the sort of detail that is implied in the Commission's draft Directive' (HOLSCEC, 1981, pp.1–2).

However, their Lordships received much more positive responses from other witnesses (though, interestingly, not the Royal Town Planning Institute: see Wood and McDonic, 1988, p.140), including a number of private developers (though the CBI remained deeply opposed). Having taken note of the Commission's attempts to accommodate the wishes of states at the drafting stage, they urged the Government to 'think European' and recognize its responsibilities to other states (who stood to benefit most from a formal EIA system) and accept the proposal on the grounds that Britain 'would not . . . have any difficulty in implementing the . . . Directive within two years' (HOLSCEC, 1981, p.xxviii).

Their report marked an important watershed in British–EU environmental relations, being the first time the Committee, which until then had forcefully criticized EU policies, had sided with the Commission against the DoE. The Committee even took the unprecedented step of holding a press conference because, as one member explained, it was 'the first [report] which departs from the [DoE] line on matters of substance' (*The Times*, 26 February 1981). The EP also tried to strengthen the draft as the negotiating proceeded in the EU, but failed (Wood, 1995, p.34). The then Chairman of its Environment Committee, Ken Collins, remembers Britain being the most forceful opponent, though France was also deeply critical:

> The UK was in the grip of an obstructive anti-Europeanism and the DoE didn't escape that. The EIA was held to be disruptive because the British planning system was so perfect that it didn't need anything from Europe. We debated constitutional practice not environmental practice and this made it difficult to get things moving.
>
> (Collins, 2000)

Having failed to achieve the necessary unanimous backing in the Council, the draft proposal returned to the Council working groups for further refinement. It was at this point that a long list of very significant changes were made to appease the British (Sheate and Macrory, 1989). These included: a reduction in the number of project types to just nine[2] (thereby making the Directive much more discretionary); a reduction in the Commission's involvement in setting thresholds under Annex II; and the deletion of a requirement to consider alternative sites (ENDS, 91, p.22; ENDS, 107, p.20; Wood, 1995, p.35). The DoE felt satisfied that it had emasculated Annex I because there were no British projects in the pipeline under at least half of the nine designated project types.

In March 1982 the Commission published a 'much modified' (Wood, 1995, p.35) version of the original proposal (CoM (82) 158 final), in the expectation that it would be accepted at the November Environment Council of that year. Having achieved most of their demands, the British withdrew their residual objections leaving the Danes to continue blocking agreement because of a domestic constitutional wrangle. The Directive was finally adopted in 1985, almost ten years after the Commission began working on a proposal.

At first, the DoE thought it had minimized Europeanization, by reducing the misfit between European and national rules. Sheate and Macrory (1989, p.73) have speculated that Britain's volte-face stemmed from the (mistaken) belief that EIAs would not be required for Annex II projects (some of which fell outside the ambit of British planning law). The ENDS Report expected the Directive's 'initial impact on planning practice in the UK . . . to be limited indeed' (ENDS, 126, p.11). Later, it explained that:

Annex I includes such projects as refineries and steelworks, which nobody is building; power stations, radioactive waste disposal sites and integrated chemical works, which will come up for approval every few years; and just one or two types of projects – such as major roads and hazardous waste disposal facilities – which will not come forward in any numbers.

(ENDS, 148, p.14)

Peter Wathern, a British expert on EIA, felt the Directive 'could have provided a vehicle for reform of the UK planning system . . . [but] what is likely to be achieved will be far more modest' (Wathern, 1988, p.207). However, some DoE planners always suspected that misfits would open up once the Directive took root in Britain (i.e., policy feedback):

The government did seem to think that . . . they had satisfactorily achieved their containment objective . . . But there was an underlying awareness of two things. First, the pressure would be irresistible in practice to have EIA for classes of project outside [Article I]. Second, in the long run it would be hard to resist application of EIA to land-use development plans and decisions on statutory instruments affecting development control [i.e., SEIA].

(Critchley, 1999)

Formal compliance

In April 1984 the DoE convened an advisory group of government officials and planners. There were two notable absentees: MAFF (which was so convinced of its immunity from EIA that it declined to send a representative!) and environmentalists, who (with the exception of the CPRE) were not invited to attend (Wathern, 1988, p.205). At an early stage, professional planners pushed for EIA to be applied across the board to all Annex I and II projects but the DoE opposed this; the line agreed across Whitehall was that EIA should *only* apply to (the limited number of) big, Annex I projects and no more (Wood and McDonic, 1988, p.14; Lambert and Wood, 1990, p.248). Two years later in 1986, the DoE issued a consultation paper and an advisory booklet. The paper encouraged the voluntary application of EIA to Annex II projects (thereby maintaining the department's strategy of tacitly supporting EIA without making it a legal requirement), but concluded that: 'the requirements of the Directive can be met within the context of the existing planning system . . . [The Government does not] foresee that it will be necessary to make carrying out of formal assessments mandatory' (DoE, 1986b, pp.2–3). At this stage, most people expected no more than half a dozen mandatory Annex I EIAs each year plus a (presumably much) smaller number of Annex II projects that had been expressly sanctioned by the SoSE (1986b, p.3).

A very strict reading of Article 4.2 seemed to justify the DoE's *de minimis* interpretation of the Directive. However, following informal representations from the Commission, the DoE was forced to accept that Annex II projects should be subject to an EIA if they were likely to generate 'significant environmental impacts'. One can imagine how badly this went down in parts of Whitehall. Apparently, officials in MAFF were furious. Like the Department of Energy, it had only assented to the Directive on the informal understanding that agricultural activities would be exempt from EIA requirements. With the threat of infringement proceedings hanging over it, the DoE set about the hard task of selling a more liberal interpretation of Annex II to the rest of Whitehall. The DoE subsequently publicized the new governmental position in a second, greatly revised, consultation paper, published in 1988 (DoE, 1988a), which represented a shift away from a case-by-case mode of implementation (supervised by the SoSE) to a more localized scheme in which local planners determined the need for EIAs according to indicative criteria drawn up centrally. Gone was any mention of 'voluntary' EIAs; projects either did or did not require an assessment according to the letter of EU law. Around this time, the DoE revised its estimate of the likely number of EIAs upwards to a couple of dozen (Wood, 1995, p.48).

Practical implementation and policy impacts

To what extent has the British government, represented by the DoE, achieved its objectives in this particular sphere of policy? In some respects, the answer is quite a lot. It wanted a fairly flexible system, covering a fairly limited range of areas and got it. It also successfully prevented the Commission from deciding which projects should undergo an EIA, and it succeeded (at least initially) in de-linking EIA from SEIA.

However, the Directive has undoubtedly had a much bigger impact in Britain than Whitehall originally intended, and generated several important unintended consequences. First and foremost, DoE Ministers originally wanted to kill the proposal outright (see HOLSCEC, 1981, p.4). As late as 1982, there were unsubstantiated reports that UKREP was calling for all negotiations to be suspended indefinitely (ENDS, 91, p.22). The proposal may well have died at this point had the HOLSCEC not taken the unprecedented step of publicly disagreeing with the DoE. Second, Britain has been forced to accept a formal and mandatory system of EIA. EIA may have been a fact of life in Britain before the EU's involvement but, left to its own devices, Britain would almost certainly have adopted a much more informal and voluntary system than the one promoted by the EU. Third, the DoE wanted to confine the scope of the Directive to a small number of large projects. In practice, far more EIAs have been undertaken than anyone expected, particularly under Annex II (Wood and Jones, 1991). Between 1988 and 1993, around 1,000 EIAs were undertaken under EU law (of which only 10 per cent were strictly required by Annex I) and over 300 outside it, indicating the tremendous societal interest generated by the Directive (Wood, 1995, p.53).

Fourth, the Directive has extended environmental protection into areas of policy that were previously exempt from planning control, such as agriculture and forestry. Moreover, the growing popularity of EIA has even encouraged developers voluntarily to undertake assessments outside the framework of EU rules. As the warning about planning delays and vexatious litigation failed to materialize, the DoE began visibly to embrace EIA with much greater enthusiasm. In 1988 the then Environment Minister, William Waldegrave, informed developers that EIA should be viewed as an opportunity 'to demonstrate to the world that [they have] done [their] environmental homework' (ENDS, 162, p.10). Today, EIA-style assessments are used in the domestic process of IPC/IPPC and in the application of the Habitats Directive. However, in spite of these changes, some parts of Whitehall remain deeply unconvinced of the merits of EIA. Thus, 16-years after the formal adoption of the Directive, MAFF had still not established legal procedures to apply EIA to certain types of environmentally damaging agricultural activity (e.g. the cultivation of semi-natural habitats; but see MAFF, 2001).

Fifth, among private actors, environmental pressure groups have, as Fiona Reynolds argues above, perhaps benefited most from the EU's involvement. For example, in a striking parallel with the Habitats Directive (Chapter 8), FoE and Greenpeace have used EU law to force the DTI to subject offshore oil exploration activities to EIA. DoE lawyers had always suspected that the Directive would apply outside territorial waters, but the Energy and Trade departments were unwilling to countenance EIA, even for presentational purposes. Environmentalists have used the experienced gained in the water sector (Chapter 7) to use the Directive to delay politically controversial road schemes and inform the Commission of lapses in compliance.

Finally, the adoption in 1997 of an amended EIA Directive suggests that the 1985 Directive has generated its own self-perpetuating policy dynamic, backed strongly by new constituencies of sub-national interest. The new Directive came about because the 1985 Directive required the Commission to submit a report on implementation after five years. Due in 1990 but not published until 1993, this report (CoM (93) 28) set in train a process of review and reform, which culminated in the Council requesting the Commission to propose an amended and improved Directive (CoM (93) 575: see Sheate, 1997b). The amending Directive (EC/97/11) strengthens the EU's EIA regime in many important respects. For example, it introduces even more formal thresholds and criteria than those in the original Directive, which have led to a doubling in the number of EIAs undertaken each year in Britain since the new Directive was implemented in 1999 (ENDS, 276, p.42; ENDS, 321, p.5). In the 1970s, the DoE had insisted that all references to criteria and thresholds be removed, because they would be 'very difficult' to draw up objectively (HOLSCEC, 1981, p.133). The 1997 Directive also introduces a number of other changes which the DoE had originally opposed in the 1980s, such as: extending the scope of EIA to include alternative sites; enlarging the list of Article I projects; and increasing public consultation. So, all in all, EIA has taken off across Europe in a way (and a form) that British planners in the DoE did their level best to avoid. To conclude, although the case-by-case paradigm of British land-use planning policy remains essentially intact, the tools used (i.e. formal EIA) and the stringency with which they are applied, have changed as a direct result of the EU involvement (see Table 11.1).

The Strategic Environmental Impact Assessment Directive

The origins of European Union policy

The Commission has coveted SEIA for almost as long as the EU has had an environmental policy. Lee and Wood had recommended it as long ago as the late 1970s, and plans to introduce it were contained in the Commission's 1980 draft EIA proposal (see above). The Commission could have used the 1997 Directive as a legal vehicle to introduce SEIA, but chose not to in case the updating process became mired in controversy. By the mid-1990s, the political context had become much more supportive of SEIA. Many states, including Britain, had begun to experiment with national-level systems of SEIA to achieve sustainable development, and the Commission was able to use the threat of inconsistent national systems to justify the need for harmonization.

The DoE's attitude to SEIA has evolved greatly over the last two decades to the point where it now regards SEIA as an important tool of sustainability policy. In the 1980s and early 1990s, Britain firmly opposed the Commission's attempts to legislate, preferring a much more informal, case-by-case approach to 'greening' national policy-making. Non-environmental departments in Britain have always presented major policy initiatives (e.g., roads, power stations) to the DoE on a 'take it or leave it' basis, and leaned heavily on the DoE to resist any incursions by the EU. The DoE's Planning Group was, in any case, philosophically opposed to a formal system of EIA, let alone SEIA, so it was happy to transmit these criticisms to Brussels. However, in the 1980s, EPG began to warm to the idea of SEIA as EIA took off in Britain (Therivel *et al.*, 1992, pp.61–2). Crucially, the EPG began to realize that the environmental appraisal of projects (i.e., SEIA) could be used to bolster EPI in Britain (see Chapter 4). Therefore, during the early 1990s it set about exploiting national channels of action to develop systems of environmental appraisal. Although these lacked the formality of SEIA, they were an attempt to place environmental limits upon policy-making in 'non' environmental sectors such as transport, energy and agriculture. The strategy of advancing SEIA 'by stealth' in Britain was formally announced in Patten's 1990 environment White Paper (Chapter 5), which contained concrete proposals for national action but resisted further EU involvement:

> EIAs impose costs both on developers . . . and on [planning] authorities . . . Any case for the extension of the application of EIA must therefore be considered carefully . . . [But] there is scope for a more systematic approach within Government to the appraisal of the environmental costs and benefits before decisions are taken . . . The aim is to provide general guidance to departments, not to set out a rigid set of procedures to be followed in all cases.
>
> (HM Government, 1990, pp.88, 231)

Environmental appraisal, or 'greening government', extended slowly and unevenly across Whitehall in the 1990s (Jordan, 2001a). However, environmental appraisal is not SEIA as conventionally understood because it only provides very limited opportunities for public involvement. This did not, however, dissuade the DoE from submitting a proposal to the Maastricht IGC, which would have forced the Commission to apply EIA/SEIA to 'green' its own proposals (see Chapter 5). Meanwhile, at the local level, the DoE produced a series of planning policy guidance

(PPG) documents that encouraged local planners to innovate in the way they appraised the environmental impacts of their long-term development plans (Wood, 1995, pp.278–80). Therivel *et al.* (1992, p.64) describe these two essentially homespun initiatives as an important 'step up' from EIA, which nonetheless lacked the formality and legal discipline of an EU system of SEIA.

The negotiation of European Union policy

Sensing that the time was now right to act, the Commission formally committed the EU to SEIA by inserting a reference into the 1987 Fourth Environmental Action Programme. It began working on a formal proposal in 1990, which was subsequently released to national experts in March 1991. Early drafts implied that SEIA would apply to just about all policies, programmes and projects that give rise to development (ENDS, 196, pp.18–20). The then environment Minister, David Trippier, condemned them as 'in truth, half-baked' (Therivel *et al.*, 1992, p.56) and told DG Environment to return to the drawing board. This was really a device to buy time because, behind the rhetoric, the DoE was caught in a cleft stick of its own making. The last thing the DoE (or any other Whitehall department) wanted to do was gift national environmental groups with another formal, legal stick but, at the same time, the EPG knew it had to improve Britain's battered environmental image in Europe. Therefore the DoE's concerns would have to be very carefully moderated so they struck the right chord. In other words, it had to think and speak more 'European' than it had in the long battle to domesticate the original EIA proposal.

Britain deployed many of the same objections it had used against EIA in the 1980s (e.g., SEIA is already implicit in national legislation; SEIA would delay the planning process, etc.), but this time they were articulated in much more *communautaire* terms (i.e., 'yes but' rather than 'no never': see Chapter 2). Consider the following example:

> We fully agree with the purpose of the proposal . . . The UK's doubts . . . do not arise because we cannot live with the basic idea. Our doubts are occasioned by the form of the proposal . . . It would require that certain [environmental] procedures be applied to all decisions . . . Policy is developing all the time [and] there is often no single moment when a decision is made . . . Obviously different governments will face these problems to different degrees, but all democratic governments will have them.
>
> (DoE, 1991)

By early 1992, the DoE had realized that a shift towards a formal, EU-level system of SEIA was probably inevitable given the groundswell of support at home and abroad, even though several other states also disagreed with the Commission about how best to proceed. Therefore, it continued to work closely with the Commission and other national environment departments to improve the proposal (which had never actually achieved formal status), rather than do what it had done in the early 1980s, and stonewall. In mid-1992, DG Environment withdrew the 'proposal', though it continued to promote SEIA through other channels (such as the provision of structural funding) and within the framework of other Directives (e.g., on habitats: Therivel *et al.*, 1992, p.53; Sheate, 1997a, pp.279–81).

Following extensive discussions with Member States, the Commission finally issued a formal proposal in December 1996 (CoM (96) 304), covering plans and programmes but not (as originally suggested) broader policies. Other changes made to appease the Member States included greater flexibility to tailor the Directive to suit national circumstances. According to one senior Commission official, British involvement during the crucial, pre-proposal stage was highly influential in delivering a more agreeable text (Hanley, 1998, p.63). By 1999, the DoE was said to be broadly in favour of the proposal although it continued to seek minor alterations to the text. Thereafter, the proposal remained in limbo as a succession of Council Presidencies (including Britain's) ignored it. Interestingly, during its 1998 Presidency of the EU, the DoE shunned SEIA in favour of EPI at the supranational level via the so-called 'Cardiff process' of review and reporting (Jordan and Lenschow, 2000). However, the 'history making' decision made at Amsterdam in 1997 (Chapter 6) to strengthen EPI in the EU gave the SEIA proposal a clearer legal justification and an all important political shot in the arm. In 1999, first the Finnish and then the German Presidencies picked up the initiative as a means of implementing Amsterdam's Committment to EPI. Ministers finally reached political agreement on the text in December 1999. It was a good measure of the DoE's ability to shape the text at the pre-Council phase that Spain was the environmental 'laggard' holding out in the Council for last minute concessions, not Britain. The text itself is significantly less ambitious than the Commission's original proposal. Successive Presidencies had whittled it down to plans and programmes in 11 specified sectors (e.g., land-use planning decisions such as local development plans and waste local plans) rather than all strategic plans and programmes. To the annoyance of environmental groups and MEPs, Ministers also exempted the EU's own development programmes from

SEIA. The Directive was finally adopted in 2001. States have until July 2004 to transpose it.

Theoretical reflection

That new land-use planning issues are still subject to unanimous voting in the Environment Council underlines their intense political sensitivity. Significantly, the EIA and SEIA Directives are the only incursions made by the EU into this most politically sensitive area of environmental policy. Yet over the course of the last 30 years, British land-use policy has been extensively and irrevocably Europeanized. The DoE was sufficiently concerned about the depth of the EU's influence to commission consultants to analyse the matter. They revealed that another important driver of Europeanization were the activities undertaken in related fields such as regional funding. The combined impact of all these EU activities 'is often indirect, and sometimes subtle' (Wilkinson, Bishop and Tewdwr-Jones, 1998, pp.22, 64).

If we turn back to consider EIA and SEIA in the context of the three key questions outlined in Chapter 1, what have been the most significant impacts in Britain? In terms of the Europeanization of policy, the EIA Directive has undoubtedly constrained the traditional, case-by-case approach to land-use planning. Britain may have been moving in the direction of EIA when the EU started to take an interest, but the EIA Directive has created a much more formal and publicly accessible system than would otherwise have emerged. It has not altered the structures of policy-making, but has brought about important and unforeseen changes in the style and the standards of British policy-making. Crucially, many of these changes would almost certainly have been resisted had the DoE known they would occur. In short, there have been important unintended consequences, generated in large part by Britain's failure to prevent a misfit appearing between EU and national policy.

With respect to the Europeanization of departmental politics, the two Directives have generated intense conflicts within Whitehall. As a direct consequence, environmental protection has been extended into areas of British environmental policy-making that used to be closed off to the DoE. In this sense, the two Directives have empowered, and will probably continue to empower, the DoE at the expense of other departments.

This brings us to the third and perhaps most puzzling aspect: the inter-departmental politics of European integration. EU land-use planning policy has, eventually, benefited EPG, but the most curious aspect of the whole 20-year saga is that the environmental parts of the DoE never consciously set out to achieve this political outcome. The DoE's ambivalent attitude to EIA in the 1970s reflected deep splits within the department. EPG was cautiously positive in view of the potential longer-term environmental benefits, whereas the Planning Group, for whom EIA was its first significant foray into EU policy-making, remained deeply – almost viscerally – opposed to any EU involvement. Consequently, as the lead department, the DoE would have communicated a very negative message to the Environment Council irrespective of any external pressure from cognate Whitehall departments. According to a former Head of the EPG's CUEP:

> [EIA was] a perfect example . . . where we could have avoided a lot of hassle if we had only tried to influence the Commission at an early stage to develop their proposal so as to go with the grain of UK practice . . . I tried to push things this way . . . but was frustrated by a Planning Directorate that was even more awkward than the Water Directorate!
>
> (Rowcliffe, 1999)

We may never know for certain whether the DoE really understood the full impact of opening the door to EIA in 1983. It is said that negotiations were conducted at a fairly low level in the Planning Group by a local government secondee. Wood (1995, p.48) speaks for many when he says that British acceptance 'may . . . have been based on a misconception'. Be that as it may, the Planning Group almost certainly did not deliberately set out to use Europe to gain leverage over other departments by home running; on the contrary, it was deeply opposed to the EU's involvement.

These findings are, of course, deeply at odds with a state-centric view of Europeanization and integration. After all, the political outcomes described above were contrary to the economic interests of many departments and incompatible with core aspects of traditional British policy and practice. They arose in spite of the efforts made by British authorities to minimize misfits by engaging in partial or selective implementation. It is very difficult to argue that the British government (or even the DoE) set out with the conscious aim of using the EU to strengthen itself by solving intractable national political problems (although it was quick to claim the political credit when public opinion became greener in the late 1980s). It spurned the opportunity to domesticate EU legislation by uploading national policies and philosophies to Europe. Consequently, the EU has continued to challenge the domestic autonomy of the British government in matters that are only tenuously connected to the logic of creating an internal market.

Process mattered. At first sight, the EIA proposal seemed relatively unthreatening, but the DoE fiercely opposed it. It extracted several important concessions but failed to anticipate the long-term consequences of delegating authority to the EU. Before long, Whitehall found itself operating within a steadily expanding framework of rules that constrained its domestic autonomy and limited future choices. There was significant policy feedback in the sense that the Europeanization triggered by the two Directives generated significant societal adaptation (e.g., the politicization of pressure groups and empowerment of the EPG), which helped extend misfits and lock the impacts (Europeanization) into place.

History also mattered. The preferences of the national actors working within these rules altered as a consequence of their involvement in European rule-making and implementation. National environmental groups began to exploit the opportunities created by the EIA Directive. In turn, societal adaptation to EU rules helped to deepen integration and extend the Europeanization of national policy. Even more noteworthy, though, was the DoE's gradual 'conversion' to EIA. Over the course of time, it has begun to realize that it can achieve higher environmental standards by exploiting European rules rather than going through national channels of action, which had traditionally been blocked by more powerful Whitehall departments. However, this was not a conscious and deliberate strategy employed by the DoE to secure leverage in Whitehall; serendipity also played an important part.

Finally, supranational agents played an active part in advancing integration and Europeanization. By working with national environmental groups, they succeeded in advancing a much more maximal interpretation of EU law than Whitehall had originally expected or desired.

11
European Integration and the Europeanization of British Environmental Governance

> We shall have full opportunity to make our views heard and our influence felt in the councils of the Community . . . The practical working of the Community . . . reflects the reality that sovereign Governments are represented round the table.
>
> (White Paper on *The UK and the European Communities*: HM Government, 1971, p.8)

The empirical chapters of this book confirm that the EU has fundamentally and irrevocably transformed the main elements of British environmental policy, namely the underlying philosophies or paradigms of action, the tools of policy and the precise setting of those tools (see Table 11.1). However, these are, as explained in Chapter 1, only the long-term manifestations of Europeanization, which the prevailing literature presents as the final products, or *outcomes*, of European integration. The broad purpose of this book is to investigate how and why these outcomes emerged in the form that they did. In other words, it tries to understand what actually *drives* Europeanization at the European level, namely the negotiation, adoption and implementation of EU environmental policies, and to analyse its long-term impact in Britain as it fits or misfits with pre-existing British institutions, practices and policies. Chapter 1 identified central government departments as the chief (though not the only) mediators (or 'fitters') of Europeanization and integration. The purpose of this final chapter is to review the seven case studies and assess the extent to which the DoE shaped, guided and steered the Europeanization of British environmental policy to suit its own (or the British Government's) purposes.

Table 11.1 Policy change in the four policy areas

	Policy paradigms	Policy tools	Instrument settings	Overall change
Bathing Water (1976)	Transformation	Transformation	Transformation	Transformation
Drinking Water (1980)	Transformation	Transformation	Transformation	Transformation
Wild Birds (1979)	Transformation	Transformation	Transformation	Transformation
Habitats (1992)	Transformation/ Accommodation	Transformation	Transformation	Transformation
AQS (1980)	Transformation	Transformation	Absorption	Accommodation
IPPC (1996)	Absorption	Absorption	Transformation/ Accommodation	Absorption/ Accommodation*
EIA (1985)	Absorption	Accommodation	Transformation	Transformation
SEIA (2001)	Absorption?*	Accommodation?*	Absorption/ Accommodation?*	Accommodation?*

*Preliminary judgement (see Chapter 1 for an explanation of key terms).

Chapter 3 introduced two contrasting theoretical perspectives on Europeanization. This chapter attempts to assess how well they capture the drivers and outcomes of Europeanization in the British environmental policy sector. It does so by returning to the three broad aspects of Europeanization described in Chapter 1, namely: (1) the DoE and the politics of European integration; (2) the DoE and the Europeanization of national policy; (3) the Europeanization of the DoE.

Understanding European integration and Europeanization

Chapter 3 identified two broad theoretical positions – state-centric and process-based – which arrive at very different predictions about the behaviour (and importance) of national departments in relation to integration and Europeanization. To the question 'When, why and how do departments *matter?*', most state-centric theorists would respond 'Not very much.' The underlying ontology of state-centric theories assumes that individual departments have little independent agency within 'the state'. They also assume that the secondary policy processes which departments govern are relatively unimportant in determining the direction and depth of integration. According to this view, departments simply align themselves to national political and economic forces. By raising the possibility of 'slack cutting', LI imbues departments with some independent agency, but desperately little empirical analysis has been undertaken to test this assertion. Otherwise, the main thrust of state-centric theories is that state executives (*not* departments) successfully manage the co-evolution of integration and Europeanization in order to maximize national economic interests and increase their autonomy by imposing their preferred policies on national groups.

The answer received from process-based theorists on the other hand, is likely to be expressed in much more conditional terms: for instance, 'Departments can matter, depending on the timing, sequence and institutional context of action.' We have seen that in certain institutional circumstances (e.g., the Environment Council) departments can make a difference by framing and implementing key pieces of secondary policy which fit their sectoral/departmental interests. However, although departments may appear to shirk inter-departmental controls, their independent actions may produce perverse or counterproductive outcomes. These arise because of the prevalence of policy feedbacks, unintended consequences and path-dependent dynamics. Therefore the underlying ontology of process-based theories is that Europeanization and integration are unpredictable and contingent processes which have inbuilt capacities to escape the grasp of (parts of) the state. Moreover, these two processes steadily transform not only the tactics of all those involved (including departments), but also their inner beliefs and political interests.

The Department of the Environment and the politics of European integration

The 'big' decisions

Chapter 3 cut into the continuing process of European integration by differentiating between 'big' and 'small' decisions. The three IGCs examined in Chapters 2, 5 and 6 were all formally inter-governmental. By and large, the British state behaved as one would expect it to behave (i.e., inter-governmentally) and there was no clear and compelling evidence that the DoE sought or achieved 'home runs' against other Whitehall departments. Rather, all three IGCs were primarily motivated not by environmental, but bigger geopolitical considerations such as economic competitiveness, security and EU reform. However, the IGCs also produced new political opportunities for national and European pressure groups. At each successive IGC, pro-environment actors such as DG Environment and a number of national environment departments (e.g. Denmark, The Netherlands, Sweden, etc.) became more and more successful at using the opportunity to embed the environmental *acquis* more deeply in the Treaties by submitting political demands, some of which (e.g., EPI; QMV) had lain dormant for years. However, the DoE either opposed (as in the case of the SEA) these environmental demands, or was forced to react to them. Even when it did attempt to upload ideas (as in the case of Maastricht and Amsterdam), it was either blocked by more powerful Whitehall departments or overruled by the core executive. To sum up, the DoE 'mattered' very little in the major Treaty changes. DG Environment was probably the chief architect of the environmental provisions of the SEA, whereas the Maastricht and Amsterdam amendments were achieved through concerted pressure from other national environment departments.

However, a state-centric view struggles to account for three crucially important aspects of the DoE's handling of the three IGCs. First, it underplays the functional linkages between one IGC and the next, which forced states to embark upon two (and with the recent (in 2000) Nice IGC, three) Treaty negotiations in politically and economically unfavourable circumstances. Environmental pressure groups did not, as LI predicts, find it difficult to mobilize effectively. In fact, at successive

IGCs, a strengthening coalition of European environmental pressure groups found more and more channels to express and achieve their demands (Stetter, 2001).

Second, it struggles to explain the precise origin of Britain's negotiating position on environmental matters. In relation to all three decisions, British negotiators found that their choices were shaped and their options limited by pre-existing EU policy commitments that had gradually accumulated in the period since the previous 'big' decision. Very often, the environmental preferences of the DoE and the core executive were endogenously determined by the Europeanizing impacts of previous pulses of European integration. In short, both history and process mattered. In fact, the IGCs were not really 'big' decisions at all: the political commitment to environmental protection was formalized first by practice, and then (and only then) by treaty decisions. In fact, with the possible exception of the 1972 Paris Summit, environment has emerged as a legitimate area of EU involvement without a 'history-making' decision having ever been taken.

Finally, the process of integration continued long after the completion of the three IGCs. In the case of the SEA, the incorporation of QMV into Environment Council business facilitated policy 'misfits' across a suite of secondary policy areas. The case study chapters reveal that the ensuing Europeanization of national policy caught the British core executive flat footed and off-guard. It had surrendered environment to EU control because it believed it was a politically and economically unimportant 'sacrifice' issue, only then to be confronted with a succession of painful policy misfits. That these outcomes arose in Britain – a hierarchical, internally coordinated and reluctant Member State – is all the more difficult for state-centric theories to explain.

To conclude, the DoE was not terribly influential in any of the IGCs, which were, in any case, not that important in shaping EU environmental policy. For these two reasons, we need to take a much closer look at the informal integration surrounding the eight 'small' decisions covered in Chapters 7–10.

The 'small' decisions

In all four areas of secondary decision-making, the DoE's response to European initiatives was often (though not always) negative, reactive and, on occasions, plainly naive. There was a widespread feeling within and outside the Department that the EU should not be prevented from pursuing environmental goals if it so wished, as long as it did not disturb the well-established conventions and systems of British policy.

However, DG Environment had rather different plans for EU and, by implication, British policy. It became highly effective at expanding EU competence in areas such as EIA, bathing and drinking water, bird protection and AQSs that really had little or nothing to do with the single market. At first, the steady expansion of the environmental *acquis* was not taken all that seriously in the national capitals, including London. Apart from a number of high-profile controversies, large parts of British industry were quite happy to leave the DoE to negotiate on their behalf (Thairs, 1998, p.156). With the sole exception of Nigel Haigh's IEEP (which, interestingly, provided both the EPG and non-governmental organizations with European expertise), British environmental pressure groups remained steadfastly 'national' in their expectations and political activities. Even as recently as the mid-1990s, nearly half of all national environmental groups in Britain had no contact whatsoever with Brussels (Lowe and Ward, 1998, p.100). In the early, formative stages of EU environmental policy, cognate Whitehall departments took a similarly relaxed view of the Commission's activities. Although the House of Lords and some professional bodies tried to raise the level of domestic political debate, the DoE negotiated many of the early environmental Directives relatively unhindered by domestic political constraints.

In theory, the 'Monnet method' of expanding the environmental *acquis* 'by stealth' (Weale *et al.*, 2000, p.9) should have provided the DoE with enormous opportunities to domesticate EU policy by uploading British policies to Brussels. As long ago as 1984, Nigel Haigh (1984, p.301) repeated the gist of the 1972 White Paper on EEC membership (see above) when he claimed that the EU provided the Department 'with a larger canvas on which to paint its own picture of what policies ought to be'. But the department never really made the most of the opportunity to advance its departmental interest at the various 'tables' of the EU. The comparatively loose inter-departmental constraints on the Department at that time permitted 'home running', but the DoE spurned the opportunity. Why was this? Domestic political factors were certainly important – British environmental policy was in the doldrums and political support for European integration was comparatively weak (Chapter 2) – but not determinant.

Process-based theories, however, do provide an additional and more credible explanation of the DoE's antipathy to EU-level policy-making, above and beyond the broader constraints of Britain's historical attitude to Europe and to the environment (see Chapter 2). First, the DoE did not have a coherent departmental view or culture in the 1970s. State-centric theorists assume that (parts of) the state have well-developed political interests. But the (newly created) DoE lacked that sense of purpose. It was certainly not an especially 'environmental' department (Chapter 2). Its expertise (and political heart) lay in technocratic affairs such as local government, urban planning and housing. These three areas inevitably attracted the best and brightest officials, and EPG remained somewhat of a backwater, dominated by scientists and environmental experts. Consequently, the DoE's high command rejected the green agenda that emanated from parts of Germany and The Netherlands in the early 1980s as being idealistic and unsophisticated.

Second, history also mattered in another important sense because the DoE was not (and never had been) a 'European' department either. Environmental issues called for supranational solutions, but the DoE felt more comfortable dealing with parochial, economic concerns such as road transport, local finance and house building.

Third, path dependence manifested itself in the DoE's robust defence of British approaches to environmental protection based on discretionary targets and localized implementation. These institutional arrangements were 'sticky' in the sense that they commanded widespread respect among policy elites of scientists and civil servants, such as Holdgate. Over a century or more, national actors had invested substantial time and resources in first creating and then adapting to these institutional arrangements, locking them firmly into place. These arrangements were perceived to 'work' well for Britain, but they were simply not uploadable to the EU. Across all four areas of policy-making, we saw how the DoE was institutionally precommitted to supporting a deeply rooted *national* environmental protection regime.

These initial pre-commitments perpetuated themselves. Thus, the EPG could have taken steps to upload the most innovative elements of national policy (e.g., nature conservation, land-use planning and air pollution), but did not, because few people in the Department expected EU policy to amount to much. Whitehall's representatives in the EU, namely the Cabinet and Foreign Offices, felt the same, and tended not to feed through advice to the EPG on how to play the Brussels game. The DoE could also have carried on vetoing EU policy, but it soon learnt that saying 'no' in Europe did not prevent misfits or thwart Europeanization. Indeed, some politicians (e.g., Howell) stood to benefit politically from saying 'yes' to particular policies, or (as in the case of Shore with the birds proposal) did not regard them as being sufficiently threatening to oppose them. So, instead, EPG tried to 'cherry pick' items from the expanding *acquis* (i.e., to embrace European integration on its own terms).

When the *acquis* began to mutate in the 1980s and touch upon more sensitive areas of national policy, the rest of Whitehall (including some of the supposedly most Europeanized departments, such as MAFF and the DTI) eventually woke up to what was happening in the environmental sector and exerted greater pressure on the DoE. Around this time, the DoE's strategy to reduce misfits was either to veto uncongenial policies in the Environment Council (e.g., the EIA Directive) or to subvert them at the implementation stage (e.g., the water and biodiversity Directives). Although Britain was not entirely negative, the DoE's main contribution during the period 1975–85 lay in what it managed to block (i.e., the development of policy at the IGCs and during the intervening periods) rather than in what it did to shape the EU in its own image.

Individual British men contributed enormously to the informal devlopment of the *acquis*, but did so from outside Whitehall (e.g., Haigh) or from within DG Environment (e.g., Fairclough, Johnson and Clinton-Davis). These and others became very adept at employing 'subterfuge' to expand the scope and stringency of European rules (Chapter 2). The case studies demonstrate some of their tactics: generating consensus by convening groups of national experts (e.g., bathing and drinking water); piggy-backing European rules on the back of international conventions (e.g., biodiversity) or marketing them as single market measures (e.g., bathing and drinking water); and sharing technical information with national and European pressure groups which function as the Commissions's 'eyes and ears' at the national level (e.g., EIA and IPPC). The expansion of the *acquis* was supported and, on occasions, bolstered greatly by ECJ rulings and support from the EP.

More than anything else, EU environmental policy developed to the extent that it did because 'non' environmental agencies, organizations and departments across Europe saw no reason to stop it. Nowhere in Europe has the environment ever been a 'show-stopper' issue (i.e., raising fundamentally important economic, political and social concerns). The popular image of the environment as a benign and unthreatening issue allowed environmentalists to insinuate environmental provisions into the European political and economic project. If there is one policy sector which exemplifies the potential for integration to occur 'between the "cracks" of the political[ly] heavyweight conferences and enlargement decisions' (Héritier, 1999, p.11), it is the environment.

Interestingly, when the DoE eventually realised that it had to be more proactive, it did not depart significantly from well-trodden institutional paths. For example, most of the ideas and policies which the DoE uploaded to Brussels in the 1990s have been of a predominantly procedural rather than substantive nature (e.g., EPI, IPPC and eco-auditing), reflecting the well-established British tradition of setting overarching policy frameworks at the national level and devolving the technicalities of implementation to local agencies (see Weale *et al.*, 2000, p.183).

To say that the DoE's change of tactics in the late 1980s was part of some long-term strategy to domesticate the EU is to overlook the extent to which the DoE consciously shunned such opportunities in the past. The case studies dramatically reveal that its reactive behaviour generated politically embarrassing and very costly misfits with national practices. The resulting political crises, coupled with structural changes to the EU, left the DoE with no other choice but to project Britain's influence in Europe. Holdgate (2000) now concedes that:

> [t]here was an underlying failure to see that . . . law making . . . would increasingly go on in Brussels . . . We missed that I'm quite sure with hindsight. And when we realised we had to give it higher priority . . . we were like the Red Queen and Alice running as fast as we possibly could to stay in the same place. But we had probably forfeited a number of opportunities by then . . . With hindsight we should have established a European environmental affairs unit in the DoE as soon as we had joined the Community, staffed with high flying civil servants . . . so we could shape European policy.

So, *when* did the DoE 'matter' most? The chapters of this book suggest that it hardly mattered at all during the big IGCs; it was simply crowded out by the core executive. And other than tempering the pro-European and pro-environment demands made by other actors, the DoE did not dramatically affect the overall course of secondary policy that much either. *How*, then, did it 'matter' most? During the IGCs it made its presence felt by either opposing the pro-integrationist plans hatched in other parts of Whitehall or by being smothered by the core executive, which has never been consistently pro-environmental or pro-European. It affected the process of informal integration most often by either vetoing or diluting pieces of legislation in order to reduce the potential misfit (and thus the extent of domestic Europeanization) with national practices.

What light do these findings shed on the theories outlined in Chapter 3? They confirm that European integration has been much more constant and 'inherently expansive' than one would expect to find in a governance system dominated entirely by states (Weale *et al.*, 2000, p.107). At several crucial stages, it has caught many if not all Whitehall departments (including the supposedly more European ones) very flat-footed and decidedly off-guard. Misfits have forced Britain to impose additional costs on powerful economic interests that had previously been immune from environmental constraints. Process-based theories usefully reveal how and why the gradual accumulation of European rules steadily constrained the autonomy of British government and sub-national economic actors, and generated serious, unexpected consequences.

State-centric theorists might be tempted to dismiss these findings as unimportant matters of 'low' politics. Well, yes, the environment did start out as an issue of 'low' politics, but since the 1970s it has metamorphosed slowly and informally to a position of much 'higher' politics, where it intrudes into many important areas of political and economic life. It is also very doubtful whether British decision-makers would have treated the environmental *acquis* quite so lightly (think of the SEA, for example) had they known that this transformation would occur.

The Department of the Environment and the Europeanization of national policy

State-centric accounts argue that states *are* capable of managing the Europeanization of national policy to achieve their national interests, whereas process-based theories point to the prevalence of surprises, unintended consequences and lapses in state control. Which theory best describes the long-term outcomes of our 'big' and 'small' decisions? The first and most obvious thing to say is that although some sub-sectors of environmental policy have been more deeply Europeanized than others, in *all* our cases the EU has altered (though not always completely transformed) the fundamental paradigms underpinning British environmental policy, the instruments used to attain policy goals and the precise setting of those instruments (Jordan, 1998b: see also Table 11.1). More specifically, it has accelerated the shift towards a more preventative policy paradigm in areas such as water and air pollution control, introduced new policy instruments such as formal AQSs and SEIA, and strengthened the setting of new and existing tools such as EIA and BPM. It has drawn attention to inconsistencies in British policy (e.g., the tendency to treat air pollution at source via emission controls, and water pollution in the receiving environment). It has also affected the general style of British policy by introducing policies with formal timetables and specific targets. British policy has become more preventative, more transparent and more legalistic as a direct result of the EU's involvement. In areas as diverse as water pollution control, air quality management, biodiversity protection and land-use planning, the EU has forced the DoE to intervene (often reluctantly) in activities that had traditionally (and happily) been devolved to front-line agencies. National environmental policy did not smoothly absorb EU environmental policy; it was transformed by it (Table 11.1).

The chapters of this book reveal all too clearly that many of these manifestations of Europeanization were neither desired nor expected either by the DoE, cognate Whitehall departments or the core executive. If anything, the (eventually mistaken) assumption that the EU would not affect Britain was nourished by the DoE itself, which used administrative Circulars rather than primary legislation to transpose Directives, and exploited every available loophole to reduce misfits while continuing to block new initiatives in the Environment Council. These tactics meant that Europeanization did not immediately show up in the formal, legislative record; analysts (like Haigh) have had to look at the *practice* of national policy to find it. According to Haigh (1999), the DoE's high command convinced itself that the 'EU effect' would be limited and short lived:

> There was a widespread assumption . . . that our great experience of pollution control would see us right whereas other countries didn't know about this. It was a shock to me to discover how the EU had affected Britain. I was looking for . . . tiny things and I wasn't expecting to find much . . . The Department doubted my findings and some flatly rejected my thesis. [The DoE] eventually realised [that Europe] mattered not because of an intellectual argument in a book by me but through force of circumstances. They were being hit on the head when their self-image was that we were on top of these things.

It was only when Europeanization grew from a series of small 'subterranean' mutations (Chapter 2) and began to assume much more politically threatening forms that the DoE began slowly and belatedly to appreciate the full potency and intrusiveness of EU environmental law. Many DoE officials interviewed during the writing of this book blamed the misfits on Britain's overzealous attitude to implementation (Weale *et al.*, 2000, pp.320–1). In fact, many of them emerged slowly, as a product of national and European pressure group activity, opportunistic Commission enforcement and a series of landmark ECJ judgments. Lax negotiation undoubtedly played a part in the DoE's acceptance of some of the early Directives (e.g., Bathing Water and possibly also EIA), but many of the most important 'misfits' were also *socially constructed* by non-state actors, who either (as in the case of the Commission) sought to recover ground lost at the negotiation stage, or (as in the case of national groups) adapted themselves to exploit the new political opportunities that the EU was creating. These opportunities included the formal site designation process (e.g., the biodiversity and water policies) and the assessment provisions of the EIA Directive. Non-state activities thus helped to realize institutional 'misfits', which would otherwise have remained latent or unexploited. In other cases, national actors adapted to the new policies, locking them in place in precisely the manner predicted by process-based theories. Think of the demand among developers for EIA for example, or the financial returns reaped by private water companies from the construction of new wastewater treatment facilities.

It is significant that the only area left relatively untouched by Europeanization – the AQS Directive (see Table 11.1) – was only very weakly integrated at the European level. In this case, the unwillingness of the Environment Council to push a maximal agenda produced – at least in Britain's case – a self-implementing piece of legislation, stripped of the policy hooks that non-state actors eagerly exploited to maximize Europeanization in other areas. Furthermore, there is little concrete evidence that the DoE purposely exploited the Europeanization of policy to gain leverage over other departments (i.e. 'home running'). With the possible exception of bathing water (specifically, Patten's 1990 'initiative' on sewage treatment) and IPPC, the transformation owes more to the political activities of non-state groups.

Some might legitimately claim that the DoE's contribution has not been entirely negative; it *has* improved the quality of European legislation by articulating a 'pragmatic' agenda which 'ensure[s] that environmental measures take full account of other interests and are not unreasonably expensive' (Sharp, 1998, p.55). There is ample evidence that other continental countries began to embrace this agenda of their own accord in the 1990s by pushing for flexible, non-regulatory and non-prescriptive EU environmental interventions, which respect the principle of subsidiarity (Jordan, 2000). Britain has also uploaded a number of domestic innovations to the EU, most notably IP(P)C. The EU is also beginning to address Britain's long-standing British concern with the *procedure* of policy (i.e., better implementation, EPI, 'lighter' forms of regulation, etc.).

Even so, it took time – and a number of rude shocks – for the DoE belatedly to recognize that it could shape the EU to suit its own interests. A former Head of EPG (1990–5), Derek Osborn (2000), recognized that:

> In the 1980s good things were being done to us against our will and we tried to partly subvert them. In the 1990s we tried to make them

work for us and reinforce the [EU's] capabilities. By then, Europe was a two-way street for us. We tried to learn from their advances as well as trying to influence them. We began to see that Europe was the more dynamic force and that things could happen on the European scale that would not be achievable in the UK in the current political climate. *Being a better European was a way of being more environmental.* So we said 'let's get with it' and help to shape European policy so as to make more environmental progress domestically as well as in the EU.

(emphasis added)

The severe political difficulties created by the Europeanization of national policy forced the DoE to reflect critically on its handling of EU business (Chapter 2). It soon realized that its officials were weak at advocating Britain's interests in the Commission's supposedly 'technical' working groups and comitology committees. It also realized the need to establish informal contacts with other actors rather than sticking to formal (i.e., intergovernmental) procedures or channels of communication (e.g., through UKREP). EPG does now try to plan ahead much more strategically (e.g., by identifying national innovations to upload to Brussels) than it did in the past, when the domestic political climate forced a more *ad hoc* and piecemeal approach to EU affairs. Finally, it spends more time trying to nurture and sustain coalitions with like-minded states. The pattern of change exhibited in Table 11.1 does suggest that these changes inside the DoE have paid off in the sense that the more recent (i.e., post 1990) Directives have had considerably less domestic impact than the early ones. However, one would need to sample many more policies conclusively to test this assertion.

What do these findings tell us about contemporary European environmental governance? The bulk of our evidence points strongly in the direction of process-based theories. At important junctures, the DoE misread the EU, failed to anticipate the long-term consequences of delegating authority or found itself arm-locked by other departments into pushing for an unsustainably narrow interpretation of Directives. The legal firewalls and opt-outs built into particularly threatening Directives failed to materialize as supranational agents pushed for – and achieved – a much more maximal interpretation of EU law, which expanded the misfit with national practices. Then, during the implementation phase, the Europeanization triggered by the Directives fostered significant societal adaptation (i.e., the politicization of pressure groups and the creation of new political opportunity structures), which helped to lock European rules firmly into place and, in turn, triggered fresh demands for deeper integration and a fresh array of domestic impacts. The feedback effects were dramatically revealed in Chapters 7 (i.e., the link between the old and revised water *acquis*) and 10 (i.e., the growing support for SEIA). The DoE struggled to understand the EU because it saw it in intergovernmental terms rather than as an expanding process of change. Fiona McConnell, formerly Head of EPINT, argues that Europeanization was potent because it unfolded slowly and insidiously as a series of small changes, with no sudden-step changes:

> There was no single critical event or moment when the penny dropped . . . [Europeanization] travelled slowly, beneath the surface like water boring through an aquifer. Had it been more obvious it would almost certainly have been resisted. It was Nigel [Haigh] really who added the first little drop of dye, but now the water is completely changed in colour as the effect has gradually dissipated.
>
> (McConnell, 1999)

The Europeanization of the Department of the Environment

The third and final aspect of Europeanization introduced in Chapter 1 was the long-term impact on the DoE itself. European integration challenges national administrative systems because most of them were designed at a time when the political centre of gravity lay inside independent sovereign states (Toonen, 1992, p.111). Some analysts claim that Whitehall has responded relatively smoothly to the challenges of EU membership. For instance, Bulmer and Burch (1998, p.604) measured Europeanization on four institutional dimensions: formal (e.g., the creation of new committees and coordination units); procedural (e.g., rules and procedures governing spending and information sharing); advisory (e.g., Cabinet Office negotiating guidelines); and cultural (e.g., the duty to inform cognate Whitehall departments). They detect substantial but gradual change along all four dimensions, which has been more or less wholly in keeping with British traditions (Bulmer and Burch, 1998, p.624). For example, there are new coordination networks linking departments, new European coordination offices or committees within each department, a proliferation of manuals and advisory booklets on how to negotiate in Europe, and new rules and institutional mechanisms for ironing out conflict between departments. However, these do not constitute a wholesale *transformation* of the state; rather, they reflect incremental alterations to well entrenched procedures and practices (Bulmer and Burch, 2000).

An alternative perspective (also outlined in Chapter 1) suggests that Europeanization and integration have fundamentally and irrevocably transformed the state through a series of complex and long-running policy feedbacks. These have brought national and supranational actors closer together and promoted common beliefs and identities. On this view, the co-evolving processes of integration and Europeanization have not only reshaped the administrative procedures and conventions of departmental life, but the very preferences and identities of its constituent parts (i.e., departments).

Which perspective offers a more plausible explanation of the DoE's relationship with Europe? A cursory inspection of the DoE suggests that the 'EU effect' has been relatively limited. The rules, procedures and conventions which govern its relationship with other actors (such as Parliament and cognate departments in Whitehall) have changed, but not dramatically so. Some new institutions have had been created to cope with the extra volume of European work (e.g., contacts with the Commission and other environment departments) but these are modelled on pre-existing patterns and traditions (see Chapter 2). EPEUR's primary task is to coordinate European matters across EPG and act as a source of information and expertise. It shadows the Environment Council and represents EPG on interdepartmental committees. It has produced new handbooks and guidance notes to improve the handling of EU business. These are the 'Bible of European business' in the department (Humphreys, 1996, p.31).

However, if we look more closely, the DoE has been shaped by the very processes that state-centric theories say it should be shaping. These internal facets include its *internal management* regime, its *tactics* and, most radically of all, its very *identity and political interests*. Taking these in turn, the EU has profoundly altered the internal workings of the department. For instance, it has altered the distribution of work in the department. Nowadays, senior Ministers regularly discuss extremely technical matters in the Environment Council where the difference between no agreement and a qualified majority can be a few milligrams of a particular substance. In the past, the Department sent scientists and relatively junior officials to Brussels to attend to the detailed content of EU legislation (Sharp, 1998, p.34). In the past, a Minister's primary responsibility was to develop the broad framework of national policy; now that task is shared with the EU.

Europe has also forced the DoE to develop new tactics to achieve its departmental interests. Neil Summerton (1999) neatly captures the different demands posed by national and European legislation:

> Because of the tyranny of the whip, if you could persuade Ministers about a particular course of action you could proceed fairly rapidly to legislation. The process of European legislation is a *lot* more complicated and a *lot* more messy. It requires a lot more effort over a much longer period with a lot less certainty about what you can achieve from the process at the end of the day. You have to convince [the SoSE], but then you have got to get down to the hard business of convincing other people in other states and the Commission that this is the direction we ought to be going in and persuade the whole convoy to go in that direction . . . It has become much more of a *negotiation* process.

There is evidence in Chapters 9 (IPPC) and 10 (SEIA) that EPG has learned to deploy more *communautaire* arguments and engage in 'corridor diplomacy' (Humphreys, 1996, p.101) to achieve its objectives. In the past, the DoE relied too heavily on the blunt stick of the national veto, but QMV requires more subtle tactics, such as coalition-building, political kite-flying (half raising an idea to gauge the reaction of other players), 'slipstreaming' (letting a negative state take the political flak for killing a bad proposal) and informal, bilateral networking outside national channels of action (1996, pp.97–101). In an attempt to improve the projection of British ideas, the DoE has also succeeded in placing more national officials in the Commission and UKREP (Hanley, 1998, p.60), and learning from their experiences by integrating them back into the Department. Finally, it has developed a more flexible approach to deal with European legislation, which involves small teams of officials tracking a particular proposal as it passes through the legislative process.

This leads to what is probably *the* most fundamental impact of all, which is that the EU has helped to make the DoE a more *environmental* department than it would otherwise have been. This change, which has occurred at the deep level of organizational values and assumptions, owes much to political crises triggered by the misfits between European and British policies. When these began to disrupt the mainstream areas of the DoE's business, Ministers were forced to be more proactive. The political crises raised the political profile of the environment across the whole of Whitehall and in the Prime Minister's office. In turn, this gave

DoE officials the impetus and authority they needed to challenge existing cultural assumptions and overcome opposing departments by working through rather than around the tight, Whitehall coordination mechanisms. In effect, the Department had to become more environmental *before* it could become more European in its outlook and activities. Insofar as there was a critical breakpoint, it was Thatcher's decision to sign the SEA:

> Thatcher's conversion to Europe made us all realise that we're part of Europe and the environment is an important policy there. It's only when we got over that hurdle . . . that the whole debate began to shift from 'does Europe have a role' to 'if it has a role what is it to be', to 'it has a role, it's quite small', to 'it has a role, it is big and we are going to influence it because we have the moral authority to do so because we're no longer the Dirty Man of Europe'.
>
> (Plowman, 2000)

Change as profound and unexpected as this, sits very oddly within a state-centric ontology.

Of course, it would be quite wrong to portray Europe as though it were the only cause of the policy and political changes described in the case study chapters. Domestic political concern, expressed through the medium of national pressure groups, would almost certainly have altered domestic policy and empowered the DoE irrespective of Europe. The steady internationalization of environmental policy-making in areas such as acid rain and ozone depletion would also have forced the DoE to come out of its shell, irrespective of the EU's involvement. And it is important not to forget that the EPG grew in size because the DoE shed many of its non-regulatory functions in the 1980s, in response to the core executive's pursuit of new public management. However, the combined impact of these domestic and international sources of change was greatly accentuated and accelerated by the quantity and legal importance of EU environmental rules.

Thinking European? Future challenges

So, has the DoE learned to think and act more 'European'? In 1998, Tony Blair ordered Whitehall to achieve a 'step change' in its relations with the rest of Europe. Apparently, some of the more domestic departments had to grope around for evidence to show that they were taking Europe seriously, but the DoE simply re-badged its European professionalism programme. Today, the Department feels a lot more comfortable working in Europe than it did before, although the journey has been a long and painful one. British politicians will doubtless continue to disagree about whether Britain should be 'in' or 'out' of Europe, but British environmental policy is so deeply Europeanized that the DoE has to treat the EU as an inextricable part of the British political system. Ministerial attitudes (e.g., Patten and Gummer) were, of course, important in the DoE's transformation, but not determinant. There were Eurosceptical SoSEs before (e.g., Shore) and after (e.g., Howard) the turning point in the late 1980s, but, whereas Shore was able to turn his back on Europe, his contemporary successors (including the arch Eurosceptic Michael Howard) have had little choice but actively to engage the Department in Europe, whether the aim was to upload policy to Brussels (e.g., IPPC), initiate policy change (e.g., the revision of the water *acquis*) or block a Commission proposal (e.g., SEIA).

To say that the DoE is better at 'thinking (and acting) European' does not necessarily mean that it always gets what it wants in Europe. To the extent that well-known pace-setter states such as Germany now detect a 'distinctly British flavour' (Wurzel, 1999, p.128) in recent EU policies such as IPPC and eco-auditing, the DoE's ability to Anglicize the EU has, as discussed above, probably improved. In 1998, the German Council of environmental advisers (the Rat von Sacverstandigen für Umweltfragen) advised that recent Directives 'increasingly [contain] design elements unknown to German environmental protection practices, but which nevertheless have to be integrated' (Rat von Sacverstandigen für Umweltfragen, 1998, p.163). Interestingly, German departments are now producing internal handbooks and guides to improve the way they operate in Brussels (Bundersministerium der Finanzen, 2000; Demmke and Unfried, 2000). Ironically, Germany now finds itself in the same position that Britain was in 20 years ago, wanting to preserve the best of national approaches while at the same time absorbing alien approaches imported from abroad.

The DoE also cannot afford to stand still; there are at least three important challenges that it will have to grapple with in the next ten years. The first is devolution. Environmental policy is one of a number of devolved issues, which mean that in future agreement will have to be reached between the DoE and the devolved assemblies and agencies in Scotland and Wales. This raises the prospect of a more complex, multi-level governance structure, akin to the one in post-war Germany. As well as complicating implementation, if the German experience is anything to go by, devolution will also greatly politicize the process through which national negotiating positions are arrived at.

The second is enlargement. Enlargement brings many new players into the frame – national environmental groups, new MEPs and national pressure groups – with whom the DoE will need to build contacts and alliances. It will also alter the dynamics of integration by introducing many new environmental laggards. In future, the DoE will probably have to readjust its attitudes and expectations as it finds itself among the environmental leaders in Europe. The shift in role will require new tactics, attitudes and alliances. And if the IPPC saga tells us anything, it is that contemporary European environmental policy-making is already so complex and unpredictable that even the most 'European' national departments struggle consistently to achieve their objectives.

Finally, there is the EP. Like most other national environment departments, the DoE used to politely ignore MEPs. However the advent of co-decision-making has forced it to engage with the Parliament. There are obvious difficulties in achieving this, not least because MEPs are European *politicians*, not civil servants. According to Ken Collins (2000):

> The DoE still has a European-intergovernmental view of the EU. MEPs are there to do its bidding and should toe the British line. They don't recognise that MEPs are members of political groups, and if they take the British line they may be in a minority. The nature of the game has changed altogether. It's all about negotiation. So the people that succeed in Brussels are not those who are good at banging the table and making their speeches. It's the people who can negotiate and do deals which is a *completely* different skill altogether. The British have never been good at this.

This takes the DoE into new and potentially difficult territory, because formally speaking civil servants remain accountable to their Ministers and, ultimately, to the British Parliament, not the EP. Somehow, all Whitehall departments (including the DoE) must find new ways of working in the new, more fluid, more multi-levelled and multi-institutional system of environmental governance that is currently emerging in Europe.

To conclude, the extent and depth of the Europeanizing impacts on the DoE are difficult to square with the modest predictions made by state-centric theories. Crucially, many of the impacts have occurred along and through unforeseen and unpredictable pathways, which the DoE has struggled to anticipate, let alone control, in the manner that state-based theories would expect. Most fundamentally of all, Europe has transformed the Department's identity and interests by raising the profile of environmental politics in Britain and imposing a framework of legal constraints on its political activities. State-centric theories argue that states consciously trade national sovereignty for greater domestic autonomy. In our case, the vagaries of integration and Europeanization made it difficult for Britain successfully to trade one off against the other. Paradoxically, Europeanization has empowered the DoE in Whitehall in spite of the department's own endeavours.

Putting the pieces together

It is a common view that the EU has Europeanized most, if not all, aspects of British environmental policy. Eurosceptics and Europhiles will continue to disagree about the EU, but there is a strikingly high level of agreement on both sides that Europeanization has benefited British environmental policy, making it stronger, more transparent and considerably broader in its overall scope, than it would otherwise have been. As Britain's chief negotiator in Brussels and the bureaucratic conduit along which most EU environmental policy flows, one would have expected the DoE to play a major part in shaping that process of change. Yet the striking finding which emerges from the case studies is that the DoE did not actually 'matter' that much. During the early stages, the DoE shunned the political opportunities which European integration created across Europe. It saw the EU as an unhelpful imposition and channelled its energies into more parochial activities. From the late 1970s onwards, the EU began slowly to Europeanize British policy, as some of the early directives unexpectedly transmogrified. The department began (belatedly) to appreciate the importance of the EU while continuing, Canute-like, to deny its influence. The DoE's failure to master the EU greatly politicized British environmental policy in ways that have, paradoxically, altered its interests and values. During the period 1990–2000, the politicization of environmental affairs generated by the unforeseen Europeanization of British policy empowered the DoE enough to win key political battles against home departments. Then (and only then) was it able to project its influence and new-found 'European' interests on to a much wider (i.e., European) plane. Until that point, the DoE lacked an effective strategy to deal with the *multi-institutional* dynamics of the EU system, which encompass the Commission and the EP as well as the Council. Because it viewed the EU through an intergovernmental lens, the channels it used tended to be *intergovernmental* (i.e., via the core executive and UKREP) and it over-looked important opportunities to network with potential allies in other governments and supranational agencies.

How, then, might we account for the unexpectedly deep Europeaniza-tion of a state with a minimalist view of European integration and a very long history of environmental concern? There are three possible answers. First, the progressiveness of British policy was always overstat-ed, not least by the British in Brussels. It is undeniably true that Britain had a well-established environmental policy prior to the EU's involve-ment, but it was primarily directed at achieving domestic policy goals and was unsuitable as a model for the rest of the EU. In all four areas of everyday policy-making, this 'first mover disadvantage' hobbled the DoE during the all-important founding moments of the European 'regu-latory competition'.

Second, Britain tried unsuccessfully to overcome 'misfits' by stopping the tide of Europeanization. Very little sustained effort was made to re-direct it, by uploading policies to Brussels. More a policy-*taker* than a policy-*maker*, the DoE channelled its resources into holding the line by vetoing proposals in the Environment Council. But, in so doing, Britain suffered the fate of those that consistently download policy from the EU: namely implementation problems, policy misfits, and performance crises (Green, Cowles Caporaso and Risse, 2000, pp.8–9). The impacts of the initial, strategic 'mistake' of trying to block Europeanization from the outside, rather than working inside Europe to modulate the path of European integration, were enduring and self-reinforcing. When com-bined with the 'stickiness' of pre-existing policy institutional forms, they also greatly reduced the scope for subsequent learning (Pierson, 2000b, pp.485, 493).

Finally, Britain did not actually resist European integration in this sec-tor as hard or as successfully as its minimalist reputation (i.e., hard nego-tiator, dutiful implementer) would imply. In fact, on occasions it either consciously (e.g., IPPC) or unconsciously (e.g., the water directives) facilitated European integration by adopting new pieces of legislation. The key reason, though, was that the core executive and the DoE's high command viewed the environment as an unimportant, 'sacrifice issue' which could be traded for more important political/economic goals, such as the single market or an opt-out in a cognate policy area. But EU envi-ronmental policy proved to be much more costly and self-perpetuating than Ministers had expected. British environmental policy was therefore Europeanized indirectly, stealthily and largely contrary to the expecta-tions of the British government.

In terms of the two theories outlined in Chapter 3, the picture which emerges strongly from the case studies is one of constant and iterative change, linking the European, national and sub-national levels of environmental governance. Thus integration generated common policies that Europeanized national political systems, which in turn altered the domestic circumstances in which national actors formed their national preferences during subsequent rounds of negotiation. The British state struggled to anticipate, let alone control, these processes and in the end was subtly transformed by them. These events can only be understood properly as a gradually unfolding *process* through which integration and Europeanization build upon and feed off one another. Consequently, British political scientists need to study European integration to explain the Europeanization of national policies; European scholars need to investigate the Europeanization of states to understand the behaviour of national actors in Europe.

The events recounted in this book are, of course, as seen through British eyes. There is every possibility that other states' experiences of the same events were entirely different (i.e., more purposeful and less surprising). The next phase of Europeanization research will need to investigate the patterns which arise when common sets of European rules come into contact with a very heterogeneous set of national administrative structures, styles and policies. But as Europeanization research seeks to become more comparative, it should not overlook the role of individual departments and ministries, for they stand at a critical position in the EU policy-making process. The British experience strongly suggests that the structure and the culture of individual departments do have an important effect on the patterns and pathways of Europeanization at the domestic level, and that this goes well beyond the political attitudes of individual ministers. For instance, departments can deliberately create misfits by uploading ambitious policy to the EU (e.g., IPPC in Britain) or they can seek to close them by subverting policies at the implementation stage. Although notions such as 'fit' and 'misfit', 'download' and 'upload', greatly simplify these processes, they need to be sensitive to the historical co-evolution of national and EU political systems. For instance, IPPC in Britain provides a case in which uploaded policies were transformed in the EU and then downloaded back to the innovator in a form that generated much more Europeanization than anyone initially suspected. The concepts of fit and misfit also lack a sense of human agency. In all our cases, many of the 'misfits' that appeared would have remained largely hypothetical had national and supranational actors not acted to give them form and social meaning. Or, as Héritier (1998) has shrewdly put it:

> Where the established policy of a Member State diverges from a clearly specified European policy mandate, there will be an expectation to adjust, which in turn constitutes a *precondition* for change. *[T]he ability to adapt will depend on the policy preferences of key actors, and the institutional reform capacity to realise policy change and to administratively adjust to European requirements.*
>
> (emphasis added)

This suggests that policy 'fit' is really just shorthand for an ongoing and highly dynamic process with many feedbacks, rather than a single, objectively definable 'one-off' event or state of being. By acting (or, most commonly, failing to act) in particular situations, the DoE played an important part in shaping the overall course of European integration in Britain. The deepest irony is that the DoE would have been a less environmental and a much less powerful department today if it had succeeded in thwarting the Europeanization of British environmental policy.

Notes

1 Learning to 'Think European'

1 A former British Ambassador to Paris and one of the chief negotiators of Britain's entry into Europe.
2 For a valuable review of various definitions, see Radaelli (2000).
3 Like Rhodes (1995, p.12), I define the core executive as 'all those organisations and procedures which coordinate central government policies, and act as final arbiters of conflict between different parts of the government machine'. This definition encompasses the Prime Minister and his/her cabinet, plus the various coordinating organizations (e.g., the Cabinet Office) and coordinating procedures. It includes Ministers in their departments, but *not* departments themselves (cf. M. Smith, 1999, p.5, who includes departments within a somewhat wider definition of the core executive).
4 According to Haigh (2001, section 1.5) *formal* compliance involves putting in place the necessary legal or administrative arrangements to give effect to EU law, whereas *practical* compliance means ensuring that the ends specified in EU Directives are actually achieved. In the lexicon of political science, these two stages correspond to policy *outputs* and *outcomes* respectively.
5 Weale *et al.* (2000, p.97) place Britain seventh on a list of 12 Member States based on their voting record in the Environment Council (i.e., for/against particular proposals). Haigh and Lanigan (1995, p.35) also place Britain in a 'median' position.
6 Management Information System for Ministers, an annual, cross-departmental process of target setting and reporting introduced by the Secretary of State, Michael Heseltine, in 1979.

2 European Union Environmental Policy and Britain

1 These section headings are based on Nigel Haigh's (1995a) abbreviated history of the DoE.

3 Theories of European Integration and Europeanization

1 I assume that Moravcsik means the core executive when he refers to 'the state'. Importantly, this interfaces directly with, but does not encompass, national departments of state, which are assumed to have their own cultures, standard operating procedures and political interests (see Chapter 1).
2 Interestingly, Golub (1997, p.20) also claims that we need to disaggregate the state to understand integration.

4 The Negotiation of the Single European Act

1 If anything, the internal market programme threatened to undermine, not support, environmental quality by generating greater economic growth – a point that even the Commission was slow to recognize and respond to (Weale and Williams, 1992).

2 Although at first the Danes and the Germans opposed QMV because they feared it would reduce their ability to apply higher environmental standards. Article 100A(4) was specifically introduced to allow them to apply higher standards if need be (Krämer, 1987, p.680).

6 The Negotiation of the Amsterdam Treaty

1 Ratification problems in Denmark, Britain and also Germany, meant that the Maastricht Treaty did not actually enter into force until November 1993.

2 See European Parliament (1996) for a summary of national positions as of January 1996.

3 The Danes tried to write EPI into each and every single article but failed (Calster and Deketelaere, 1998, p.18).

7 Water Policy

1 Demmke (1994, p.131) claims that the two proposals were brokered by virtually the same group of national scientific experts and Commission officials.

2 It is widely believed that the standards were based upon American research conducted in the early 1970s.

3 The NWC advised Ministers and acted as a spokesbody for the industry.

4 See Chapter 1 for a definition of formal and practical compliance.

5 In 1996, the additional compliance costs in the UK were estimated to be up to £9,800 million and £20–4,250 million for drinking and bathing water respectively (Jordan, 1999).

8 Biodiversity Policy

1 The most significant are the Ramsar (1971) and Bern Conventions (1979), which deal with wetlands and habitats respectively.

2 Interestingly, the DoE assumed (incorrectly, as it turned out) that, without prior protection under British law, the Directive would be toothless.

3 In the Leybucht case, the ECJ rejected the Commission's argument that the protection of SPAs was an absolute duty other than when there were risks to human life. But it also rejected the German Government's demand for a wide margin of discretion when identifying SPAs. The Habitats Directive responds to these fears by permitting states to take social and economic factors into account when managing SPAs.

4 This term is not defined in EU law. However, it amounts to the offer of alternative habitat of a similar size and ecological quality to that lost, or the restoration of the original habitat (Nolkaemper, 1997).

5 Article 7 states that the obligations under Articles 6 (2) (3) (4) of the Habitats Directive should replace obligations arising under the first sentence of 4(4) of the Birds Directive.

6 Neofunctionalists would presumably interpret the functional link between the two Directives as an example of technical spillover.

9 Air Policy

1 Remember that DG Environment was not created until 1981 (see Chapter 2).

2 A Commons Standing Committee asked the DoE for a report on the potential impacts of the proposal in February 1994, but this was not submitted until 2 June 1995 when negotiations in the Council were nearing completion. The DoE later accepted that it had incorrectly lifted a Parliamentary scrutiny reserve (a temporary delay on EU negotiations to allow national parliamentary scrutiny) on the IPPC negotiations in order to reach a common position in the Environment Council on 24 June 1995.

10 Land-Use Planning Policy

1 Apparently, the MoD voluntarily produced an EIA for the Faslane navel base which complied with the very Commission proposal that the DoE was trying to emasculate in Brussels (Wathern, 1988, p.199)!

2 The 1980 proposal contained a list of 35 (ENDS, No.91, p.22).

Bibliography

Adams, W. (1986) *Nature's Place* (London: Allen & Unwin).

Armstrong, K. and S. Bulmer (1998) *The Governance of the Single European Market* (Manchester: Manchester University Press).

Ashby, E. (1972) *Polllution: Nuisance or Nemesis?* (London: HMSO).

Ashby, E. and M. Anderson (1981) *The Politics of Clean Air* (Oxford: Clarendon Press).

Aspinwall, M. (2000) 'Structuring Europe', *Political Studies*, 48 (3), 415–42.

Aspinwall, M. and G. Schneider (2000) 'Same Menu, Separate Tables', *European Journal of Political Science*, 38 (1), 1–36.

Bache, I. and A. McGillivray (1997) 'Testing the Extended Gatekeeper' in J. Holder (ed.), *The Impact of EC Environmental Law in the UK* (Chichester: John Wiley).

Baker, D., A. Gamble and S. Ludlam (1993) 'Whips or Scorpions? The Maastricht Vote and the Conservative Party', *Parliamentary Affairs*, 46 (2), 151–66.

Ball, S. (1997) 'Has Britain Implemented the Habitats Directive Properly?' in J. Holder (ed.), *The Impact of EC Environmental Law in the UK* (Chichester: John Wiley).

Barnett, J. (1982) *Inside the Treasury* (London: André Deutsch).

Baun, M. (1995) 'The Maastricht Treaty as High Politics', *Political Science Quarterly*, 110 (4), 605–24.

Bender, B. (1991) 'Whitehall, Central Government and 1992', *Public Policy and Administration*, 6, 13–20.

Blair, A. (1999) *Dealing with Europe: Britain and the Negotiation of the Maastricht Treaty* (Aldershot: Ashgate).

Bomberg, E. and J. Peterson (1993) 'Prevention from Above? The Role of the EC' in M. Mills (ed.), *Prevention, Health and British Politics* (Aldershot: Avebury).

Börzel, T. (1999) 'Institutional Adaptation to Europeanisation in Germany and Spain', *Journal of Common Market Studies*, 37 (4), 573–96.

Börzel, T. and T. Risse (2000) *When Europe Hits Home: Europeanisation and Domestic Change*, Robert Schumann Centre for Advanced Studies Working Paper RSC 2000/56 (Florence: European University Institute).

Brewin, C. and R. McAllister (1991) 'Annual Review of the Activities of the EC in 1990', *Journal of Common Market Studies*, 29 (4), 385–430.

Brown, K. (1993) 'Biodiversity' in D. Pearce *et al.*, *Blueprint Three* (London: Earthscan).

Budden, P. (1994) 'The Making of the Single European Act', unpublished D.Phil. Thesis (Oxford: University of Oxford).

Buller, J. and M. Smith (1998) 'Civil Service Attitudes Towards the European Union', in D. Baker and D. Seawright (eds), *Britain For and Against Europe* (Oxford: Clarendon Press).

Bulmer, S. and M. Burch (1998) 'Organising for Europe', *Public Administration*, 76, 601–28.

Bulmer, S. and M. Burch (2000) 'Coming to Terms With Europe', Paper for the University Association for Contemporary European Studies (UACES) Conference,

Budapest, April 2000. Revised version available at: http://www. qub.ac.uk/ies/onlinepapers/poe.html

Bundersministerium der Finanzen (2000) *EU: Handbuck* (Berlin: Bundersministerium der Finanzen).

Bungarten, H. H. (1978) *Umweltpolitk in Westeuropa* (Bonn: Europa Union Verlag).

Burch, M. and I. Holliday (1996) *The British Cabinet System* (London: Harvester Wheatsheaf).

Butler, D., A. Adonis and T. Travers (1994) *Failure in British Government: The Politics of the Poll Tax* (Oxford: Oxford University Press).

Byatt, I. (1996) 'The Impact of Water Directives on Water Customers in England and Wales', *Journal of European Public Policy*, 3 (4), 665–74.

Cabinet Office (1999) *The Guide To Better European Regulation*, Cabinet Office Regulatory Impact Unit (London: Cabinet Office).

Calster, G. and K. Deketelaere (1998) 'Amsterdam, the IGC and Greening the Treaty', *European Environmental Law Review*, January, 12–25.

CEC (1984) *Ten Years of Community Environment Policy* (Brussels: CEC).

CEC (1987) *Fourth Environmental Action Programme*, 7 December 1987 (Brussels: CEC).

Christiansen, T., K.-E. Jorgensen and A. Weiner (eds) (1999) 'The Social Construction of Europe', *Journal of European Public Policy*, 6 (4), 527–718.

Christoph, J. (1993) 'The Effect of Britons in Brussels', *Governance*, 6 (4), 518–37.

Climate Network Europe, *et al.*, (1995) *Greening the Treaty II: Sustainable Development in a Democratic Union: Proposals for the 1996 IGC* (Utrecht, The Netherlands: Stichting Natuur en Milieu).

Clinton-Davis, S. (1996) Interview with the author, London, 4 July 1996.

Collins, K. (2000) Interview with the author, London, 23 May 2000.

CoM (1995) *Reflection Group's Report*, SN 520/95, 5–12–95 (Brussels: CoM).

Corbett, R. (1987) 'The 1985 IGC and the SEA' in R. Pryce (ed.), *The Dynamics of the EU* (London: Croom Helm).

Corbett, R. (1992) 'The IGC on Political Union', *Journal of Common Market Studies*, 30 (3), 271–98.

Corbett, R. (1997) 'Governance and Institutional Developments', *Journal of Common Market Studies*, 35, Annual Review, 37–51.

Council of the European Communities (1995) *Progress Report from the Chairman of the Reflection Group on the IGC 1996*, SN 509/1/95, 1 September 1995 (Brussels: CoM).

Council of the European Communities (1996) *The European Union Today and Tomorrow: 'General Outline for a Draft Revision of the Treaties'*, Brussels, 5 December (Brussels: CoM).

Cox, C., P. Lowe and M. Winter (1986) 'Agriculture and Conservation in Britain' in C. Cox, P. Lowe and M. Winter (eds), *Agriculture* (London: Allen & Unwin).

CPRE and the Green Alliance (1995) *Greening the Treaty: A Manifesto for the IGC from UK Member of the EEB* (London: CPRE/Green Alliance).

Cram, L. (1997) *Policy Making in the EU* (London: Routledge).

Critchley, P. (1999) Correspondence with the author, 19 November 1999.

de Ruyt, J. (1987) *L'Acte Unique Européen* (Brussels: Editions de l'Université de Bruxelles).

Dehousse, F. (1999) 'The IGC Process and Results' in P. O'Keefe and P. Twomey (eds), *Legal Issues of the Amsterdam Treaty* (Oxford: Hart).

Dehousse, R. (1998) *The European Court of Justice* (London: Macmillan – now Palgrave Macmillan).

Dehousse, R. and Majone, G. (1989) 'The Institutional Dynamics of European Integration' in S. Martin (ed.), *The Construction of Europe* (Dordrecht, Netherlands: Kluwer).

Demmke, C. (1994) *Die Implementation der EG-Umweltpolitik in den Mitgliedstaaten* (Baden Baden: Nomos).

Demmke, C. and M. Unfried (2000) *Umweltpolitik Zwishen Brussel und Berlin: Ein Leitfaden für die Deutsch Unweltverwaltung* (Maastricht: European Institute of Public Administration).

Demmke, C. and M. Unfried (2001) *European Environmental Policy: The Administrative Challenge for Member States* (Maastricht: European Institute for Public Administration).

DETR (1997) *The UK National Air Quality Strategy*, Cmnd 3587 (London: DETR).

DETR (1999) *UK Smoke and Sulphur Dioxide Monitoring Network – Summary Tables for April 1997–March 1998* (London: DETR). http://www.aeat.co.uk/ netcen/ airqual/reports/

DETR (2000a) *Countryside Survey 2000* (London: DETR) http://www.wildlife-countryside.defra.gov.uk/cs2000/index.htm

DETR (2000b) *The Air Quality Strategy for England, Scotland, Wales and Northern Ireland*, Cmnd 4548 (London: The Stationery Office).

Devuyst, Y. (1999) 'The Community Method After Amsterdam', *Journal of Common Market Studies*, 37 (1), 109–20.

Dinan, D. (1994) *Ever Closer Union* (London: Macmillan – now Palgrave Macmillan).

Dinan, D. (1997) 'The Commission and the Reform Process', in G. Edwards and A. Pijpers (eds), *The Politics of European Treaty Reform: The 1996 IGC and Beyond* (London: Pinter).

Dinan, D. (1998) 'Reflections on the IGCs' in P. Laurent and M. Maresceau (eds), *The State of the European Union: Volume 4* (Boulder, CO: Lynne Rienner).

Dinan, D. (1999) *Ever Closer Union (2e)* (London: Macmillan – now Palgrave Macmillan).

Dixon, J. (1998) 'Nature Conservation' in P. Lowe and S. Ward (eds), *British Environmental Policy and Europe* (London: Routledge).

DoE (1972) *The Human Environment: The British View* (London: DoE).

DoE (1978a) *Pollution Control in Great Britain: How It Works*, DoE CUEP Pollution Paper (London: HMSO).

DoE (1978b) Letter from Peter Shore to the EEB/Nigel Haigh dated 24 May *SH/PSO/12513/78* (London: DoE).

DoE (1978c) 'Peter Shore on Planning Inquiries', *DoE Press Notice No. 488*, 13 September (London: DoE).

DoE (1979) *Advice on the Implementation in England and Wales of the EEC Directive on the Quality of Bathing Water*, 9 July (London: DoE).

DoE (1980a) *Letter from Miss Lillian Birch, 8 January 1980 to all interested groups, Preliminary Draft Directive, EIA/OU/18* (London: DoE).

DoE (1980b) *MINIS 2* (London: DoE).

DoE (1981) *Circular 11/81 Clean Air* (London: DoE).

DoE (1982a) *EC Directive Relating to Quality of Water Intended for Human Consumption (80/779/EEC)*, DoE Circular 20/82 (London: DoE).

DoE (1982b) *The Information and Standstill Agreement*, E/G/81–82: 54, 25 March (London: DoE).

DoE (1986a) *Air Pollution Control in Great Britain: A Consultation Paper*, December (London: DoE).

DoE (1986b) *Implementation of the European Directive on EIA*, Consultation Paper (London: DoE).

DoE (1986c) *MINIS 7, Part 4, Environmental Services and Water*, July (London: DoE).

DoE (1987) *MINIS 8, Part 4* (London: DoE).

DoE (1988a) *Environmental Assessment: Implementation of the EC Directive*, Consultation Paper (London: DoE).

DoE, (1988b) *MINIS 9, Part 4* (London: DoE).

DoE (1989) *MINIS 10, Part 4* (London: DoE).

DoE (1990) *MINIS 11, Water Directorate* (London: DoE).

DoE (1991) 'Strategic Environmental Assessment', *DoE News Release*, 28 October (London: DoE).

DoE (1993a) *Explanatory Memorandum on IPPC Framework Directive, 9491/93, ENV 292*, 12 November (London: DoE).

DoE (1993b) *MINIS 14, Part 4* (London: DoE).

DoE (1993c) *The Links Between the DoE and the Institutions of the EC ('The Bristow Report')*, December (London: DoE).

DoE (1995) *Briefing: Proposed Council Directive on IPPC from David Mottershead to Caroline Jackson and all UK MEPs*, 11 October (London: DoE).

Draper, P. (1977) *Creation of the DoE* (London: HMSO).

Drewry, G. (1995) 'The Case of the UK' in S. Pappas (ed.), *National Administrative Procedures for the Preparation and Implementation of Community Decisions* (Maastricht: European Institute of Public Administration).

Drinking Water Inspectorate (1999) *Drinking Water Quality: Tenth Annual Report* (London: Drinking Water Inspectorate).

DTI (1993) *Review of the Implementation and Enforcement of EC Law* (London: DTI).

Edwards, G. (1992) 'Central Government' in S. George (ed.), *Britain and the EC* (Oxford: Clarendon Press).

Edwards, G. and A. Pijpers (eds) (1997) *The Politics of European Treaty Reform: The 1996 IGC and Beyond* (London: Pinter).

Emmott, N and N. Haigh (1996) 'Integrated Pollution Prevention and Control', *Journal of Environmental Law*, 8 (2), 301–11.

Environmental Data Services Ltd (ENDS), *ENDS Report*, various issues.

ENDS Daily (1997) 'IGC Talks Focus on "Environmental Guarantee"', *ENDS Daily* 30 May.

Environment Watch: Western Europe (1995a) 'Place of Environment Uncertain as EU Treaty Debate Starts', *Environment Watch: Western Europe*, 2 June, p.11.

Environment Watch: Western Europe (1995b) 'High Level Group Identifies Environment as Priority for EU Treaty Reform', *Environment Watch: Western Europe*, 15 September, p.17.

Environment Watch: Western Europe (1996) 'Commission Proposes Bolstering Environment Dimension of EU Treaty', *Environment Watch: Western Europe*, 1 March, pp.11–12.

European Parliament (Committee on the Environment, Public Health and Consumer Protection) (1988) *First Working Document on the Application of the Birds Directive in the EC*, March (Brussels: European Parliament).

European Parliament (1996) *White Paper on the 1996 IGC, Volume II: Summary of Positions of the Member States of the EU With a View to the 1996 IGC* (Strasbourg: European Parliament, IGC Taskforce), *http://www.europa.eu.int/en/agenda/igc-home/*

Evans, D. (1973) *Britain in the EC* (London: Victor Gollancz).

Evans, P. (1980) *EC Directive on Smoke and Sulphur Dioxide*, National Society for Clean Air (NSCA) 47th Annual Conference, 22–25 September.

Fairclough, A. (1999) Interview with the author, London, 8 December.

FCO (Foreign and Commonwealth Office) (1996) *A Partnership of Nations: The British Approach to the EU IGC 1996*, Cmnd 3181, March (London: HMSO).

Forster, A. (1998) 'Britain and the Negotiation of the Maastricht Treaty', *Journal of Common Market Studies*, 36 (3), 347–68.

Forster, A. (1999) *Britain and the Maastricht Negotiations* (London: Macmillan – now Palgrave Macmillan).

Francis, J. (1994) 'Nature Conservation and the Voluntary Principle', *Environmental Values*, 3 (3), 267–72.

Freestone, D. (1996) 'The Enforcement of the Wild Birds Directive' in H. Somsen (ed.), *Protecting the European Environment* (London: Blackstone Press).

Gameson, A. (1979) 'EEC Directive on Quality of Bathing Water', *Water Pollution Control*, 80 (2), 221–31.

Gammell, A. (1987) *Manual on European Council Directives on the Conservation of Wild Birds* (Brussels: EEB).

Geddes, A. (1994) 'Implementation of Community Environmental Law: Bathing Water', *Journal of Environmental Law*, 6 (1), 125–35.

George, S. (1994) *An Awkward Partner (2e)* (Oxford: Oxford University Press).

George, S. (1996) 'The Approach of the British Government to the 1996 IGC', *Journal of European Public Policy*, 3 (1), 45–62.

Glasson, J., R. Therivel and A. Chadwick (1994) *Introduction to EIA* (London: UCL Press).

Golub, J. (1996) 'British Sovereignty and the Development of EC Environmental Policy', *Environmental Politics*, 5 (4), 700–28.

Golub, J. (1997) 'The Path To EU Environmental Policy', Paper presented at the Fifth Biennial International Conference of the European Community Studies Association ECSA, Seattle, May.

Gooriah, B and M. Williams (1982) 'Progress in Meeting the EC Directive With Specific Reference to Smoke Control', Paper presented at the NSCA 49th Annual Conference, 19 October.

Gowland, D. and A. Turner (2000) *Reluctant Europeans* (New York: Addison-Wesley-Longman).

Grant, C. (1994) *Delors: Inside the House that Jacques Built* (London: Nicholas Brearley).

Green Alliance (1995) *Greening the Treaty: Notes of a CPRE/Green Alliance Seminar* (London: Green Alliance).

Green Cowles, M., J. Caporaso and T. Risse (eds) (2000) *Transforming Europe* (Ithaca, NY: Cornell University Press).

Griller, S. *et al.* (2000) *The Treaty of Amsterdam: Facts, Analysis, Prospects* (Vienna: Springer Verlag).

Gummer, J. (2000) Interview with the author, London, 6 March.

Haigh, N. (1983) 'The EEC Environmental Assessment Directive', Paper presented at the International Land Reclamation Conference, Grays, Essex, 26–9 April.

Haigh N. (1984) *EEC Environmental Policy and Europe* (London: ENDS).

Haigh, N. (1989a) *Manual of Environmental Policy*, 2nd rev. edn (Harlow: Longman).

Haigh, N. (1989b) *Possibilities for the Development of a Community Strategy on Integrated (Multi-Media) Pollution Control* (London: IEEP).

Haigh, N. (1992) *Manual of Environmental Policy* (London: Carter Mill).

Haigh, N. (1995a) 'Environmental Protection in the DoE' in DoE (ed.), *A Perspective For Change* (London: DoE).

Haigh, N. (1995b) 'Initial Thoughts on a Paper from the Netherlands Dated May 1995, and Headed "Environmental Framework Directives of the European Union" ', Mimeo dated 22 August (London: IEEP).

Haigh, N. (1998) 'Introducing the Concept of Sustainable Development into the Treaties of the EU' in T. O'Riordan and H. Voisey (eds), *The Transition to Sustainability* (London: Earthscan).

Haigh, N. (1999) Interview with the author, London, 29 April.

Haigh, N. (ed.) (2001) *Manual of Environmental Policy* (London: IEEP/Elsevier).

Haigh, N. and D. Baldock (1989) *Environmental Policy and 1992* (London: IEEP).

Haigh, N and F. Irwin (eds) (1989) *Integrated Pollution Control in Europe and North America* (London and Washington: Conservation Foundation and IEEP).

Haigh, N. and C. Lanigan (1995) 'Impact of the European Union on UK Environmental Policy Making' in T. Gray (ed.), *UK Environmental Policy in the 1990s* (London: Macmillan – now Palgrave Macmillan).

Hajer, M. (1995) *The Politics of Environmental Discourse* (Oxford: Oxford University Press).

Hall, P. (1993) 'Policy Paradigms, Social Learning and the State', *Comparative Politics*, 25 (3), 275–96.

Hall, T. (1999) Correspondence with the author, 30 November 1999.

Hanley, N. (1998) 'Britain and the European Policy Process' in P. Lowe and S. Ward (eds), *British Environmental Policy and Europe* (London: Routledge).

Hannay, D. (ed.) (2000) *Britain's Entry into the EC* (London: Frank Cass).

Hatton, C. (2000) Interview with J. Fairbrass, Godalming, 26 July.

Hayes-Renshaw, F. and H. Wallace (1996) *The Council of Ministers* (London: Macmillan – now Palgrave Macmillan).

Heath, E. (1998) *The Course of My Life* (London: Hodder & Stoughton).

Heclo, H. and A. Wildavsky (1981) *The Private Government of Public Money*, 2nd edn (London: Macmillan – now Palgrave Macmillan).

Hepburn, I. (2000) Interview with J. Fairbrass, Redgrave, 18 July.

Héritier, A. (1998) 'Differential Europe: National Administrative Responses to Community Policy', Mimeograph (Florence: European University Institute).

Héritier, A. (1999) *Policy Making and Diversity in Europe* (Cambridge: Cambridge University Press).

Héritier, A. *et al.* (1996) *Ringing the Changes* (Berlin: De Gruyter).

Hill, M., S. Aaronovitch, and D. Baldock (1989) 'Non-Decision Making in Pollution Control in Britain', *Policy and Politics*, 17 (3), 227–40.

HM Government (1971) *The UK and the European Communities*, White Paper (London: HMSO).

HM Government (1984) 'Europe – the Future', *Journal of Common Market Studies*, 23 (1), 73–81.

HM Government (1990) *This Common Inheritance* (London: HMSO).

HM Government (1994) *Sustainable Development*, Cmnd 2426 (London: HMSO).

HM Government, (1996) *Intergovernmental Conference: Quality of Legislation – Memorandum by the UK, July 1996* (London: FCO): http://www.europa.eu.int/en /agenda/igc-home/ms- doc/state-uk/qualleg.htm.

HOCESCA (1995) *Integrated Pollution Prevention and Control*, 13 December 1995 (London: HMSO).

HOCSCSL (1975) *Minutes of Evidence Taken Before the Select Committee, 1 July 1975*, Session 1974–5, HC 87- viii (London: HMSO).

HOCSCSL (1977) *Bird Conservation*, Twenty Fifth Report, Session 1976–7 (London: HMSO).

Hogg, S. and J. Hill (1995) *Too Close to Call* (London: Little, Brown).

Holdgate, M. (1979) *A Perspective on Pollution* (Cambridge: Cambridge University Press).

Holdgate, M. (1983) 'Environmental Policies in Britain and Mainland Europe' in R. Macrory (ed.), *Britain, Europe and the Environment* (London: ICCET, Imperial College London).

Holdgate, M. (2000) Interview with the author, Trumpington, 30 March.

HOLSCEC (1975) *Thirtieth Report*, Session 1974–5, HL Paper 298 (London: HMSO).

HOLSCEC (1976a) *Nineteenth Report*, Session 1975–6, HL Paper 111 (London: HMSO).

HOLSCEC (1976b) *R/90/76 and R/540/76: Sulphur Pollution*, HL 309 (London: HMSO).

HOLSCEC (1978) *Approximation of Laws Under A100 of the EEC Treaty*, 22nd Report, HL 131 (London: HMSO).

HOLSCEC (1979) *Environmental Problems and the Treaty of Rome* (London: HMSO).

HOLSCEC (1981) *Environmental Assessment of Projects*, 11th Report, Session 1980–1 (London: HMSO).

HOLSCEC (1987) *Fourth Environmental Action Programme*, Eighth Report, HL 135, Session 1986–7 (London: HMSO).

HOLSCEC (1989) *Habitats and Species Protection*, Fifteenth Report, Session 1988–89 (London: HMSO).

HOLSCEC (1994) *Bathing Water – With Evidence*, Session 1994–5, First Report, HL Paper 6-I (London: HMSO).

HOLSCEC (1995) *The 1996 IGC:* 21st Report, HL Paper 105 (London: HMSO).

HOLSCEC (1996) *Drinking Water*, Session 1995–6, Fourth Report, HL Paper 31 (London: HMSO).

HOLSCEC (1999) *Biodiversity in the EU: Interim Report*, Eighteenth Report, Session 1998–99 (London: HMSO).

Hood, C. (1995) 'De-Privileging the UK Civil Service' in J. Pierre (ed.), *Bureaucracy in the Modern State* (London: Edward Elgar).

Hovden, E. (2002) 'EU Environmental Policy and the Choice of Legal Base', *Environment and Planning C* (in press).

Howe, G. (1994) *A Conflict of Loyalty* (London: Macmillan – now Palgrave Macmillan).

Howell, D. (1975) *The Environment and the European Communities*, Text of a Speech, Birmingham, April.

Humphreys, J. (1996) *A Way Through the Woods* (London: DoE).

IEEP (1995) *The 1996 IGC: Integrating the Environment into other EU Policies* (London: IEEP).

International Environmental Reporter (1995) 'Conflict on the Role of Environment at Meeting on Maastricht Seems Destined', *International Environmental Reporter*, 6 September, p.671.

Johnson, S. (2000) Interview with the author, London, 8 May.

Jordan, A. J. (1993) 'Integrated Pollution Control and the Evolving Style and Structure of Environmental Regulation in the UK', *Environmental Politics*, 2 (3), 405–27.

Jordan, A. J. (1997) ' "Post-Decisional" Politics in the EC', unpublished PhD thesis (Norwich: University of East Anglia).

Jordan, A. J. (1998a) 'Step Change or Stasis? EC Environmental Policy After the Amsterdam Treaty', *Environmental Politics*, 7 (1), 227–36.

Jordan, A. J. (1998b) 'The Impact on UK Environmental Administration' in P. Lowe and S. Ward (eds), *British Environmental Policy and Europe* (London: Routledge).

Jordan, A. J. (1998c) 'Private Affluence and Public Squalor? The Europeanisation of British Coastal Bathing Water Policy', *Policy and Politics*, 26 (1), 33–54.

Jordan, A. J. (1999) 'European Community Water Policy Standards', *Journal of Common Market Studies*, 37 (1), 13–37.

Jordan, A. J. (2000) 'The Politics of Multilevel Environmental Governance: Subsidiarity and Environmental Policy in the European Union', *Environment and Planning A*, 32 (7), 1307–24.

Jordan, A. J. (2001a) 'Efficient Hardware, Light Green Software? EPI in the UK' in A. Lenschow (ed.), *Environmental Policy Integration: Greening Sectoral Policies in Europe* (London: Earthscan).

Jordan, A. J. (2001b) 'National Environmental Ministries: Managers or Ciphers of European Environmental Policy?', *Public Administration*, 79 (3), 643–63.

Jordan, A. J. (2001c) 'The European Union: An Evolving System of Multi-Level Governance . . . or Government?', *Policy and Politics*, 29 (2), 193–208.

Jordan, A. J. (2003) 'The Europeanization of British Government and policy', *British Journal of Political Science* (in press).

Jordan, A. J., R. Brouwer and E. Noble (1999) 'Innovative and Responsive? A Longitudinal Analysis of the Speed of EU Environmental Policy Making', *Journal of European Public Policy*, 6 (3), 376–98.

Jordan, A. J. and J. Fairbrass (2002) 'EU Environmental Policy After the Nice Summit', *Environmental Politics,* 10,4,109–114 (in press).

Jordan, A. J. and J. Greenaway (1998) 'Shifting Agendas, Changing Regulatory Structures and the "New" Politics of Environmental Pollution', *Public Administration*, 76, 669–94.

Jordan, A. J. and T. Jeppesen (2000) 'EU Environmental Policy: Adapting to the Principle of Subsidiarity?', *European Environment*, 10 (2), 64–74.

Jordan, A. J. and A. Lenschow (2000) ' "Greening" the European Union', *European Environment*, 10 (3), 109–20.

Judge, D. (1985) 'The British Government, European Union and EC Institutional Reform', *Political Quarterly*, 57 (3), 321–8.

Kassim, H., B. G. Peters and V. Wright *et al.* (eds) (2000) *The National Coordination of EU Policy: The Domestic Level* (Oxford: Oxford University Press).

Keohane, R., B. King and S. Verba (eds) (1994) *Designing Social Inquiry* (Princeton, NJ: Princeton University Press).

Keohane, R. and J. Nye (1972) *Transnational Relations and World Politics* (Cambridge, MA: Harvard University Press).

Keohane, R. and J. Nye (1989) *Power and Interdependence*, 2nd edn (Glenview, IL: Scott, Foresman).

Knill, C. (1997) 'The Europeanisation of Domestic Policies', *Environmental Policy and Law*, 27 (1), 48–57.

Knill, C. (2001) *The Europeanisation of National Administrations* (Cambridge: Cambridge University Press).

Köppen, I. (1993) 'The Role of the ECJ' in D. Liefferink *et al.* (eds), *European Integration and Environmental Policy* (London: John Wiley).

Krämer, L. (1987) 'The Single European Act and Environmental Protection', *Common Market Law Review*, 24, 659–88.

Lambert, A. and C. Wood (1990) 'UK Implementation of the European EIA Directive', *Town Planning Review*, 61 (3), 247–61.

Langrish, S. (1998) 'The Treaty of Amsterdam', *European Law Review*, 23 (1), 3–20.

Laursen, F. (1992) 'Explaining the IGC on Political Union' in F. Laursen and S. Vanhoonacker (eds), *The Intergovernmental Conference on Political Union* (Maastricht: Martinus Nijhoff).

Laursen, F. and S. Vanhoonacker (eds) (1992) *The Intergovernmental Conference on Political Union* (Maastricht: Martinus Nijhoff).

Ledoux, L., S. Crooks, A. Jordan and R. Turner (2000) 'Implementing European Union Biodiversity Policy: UK Experiences', *Land Use Policy*, 17 (4), 257–68.

Lee, N. and C. Wood (1978) 'EIA of Projects in EEC Countries', *Journal of Environmental Management*, 6, 57–71.

Levitt, R. (1980) *Implementing Public Policy* (London: Croom Helm).

Liefferink, D. (1996) *Environment and the Nation State* (Manchester: Manchester University Press).

Lindberg, L. and S. Scheingold (1970) *Europe's Would-be Polity* (Englewood Cliffs, NJ: Prentice-Hall).

Lodge, J. (1986) 'The SEA: Towards a New Euro-Dynamism?', *Journal of Common Market Studies*, 24 (3), 203–23.

Long, T. (1998) 'The Environmental Lobby' in P. Lowe and S. Ward (eds), *British Environmental Policy and Europe* (London: Routledge).

Lord, C. and N. Winn (1998) 'Garbage Cans or Rational Decisions? Member Governments, Supranational Factors and the Shaping of the Agenda for the IGC', Mimeo (Leeds: University of Leeds).

Lowe, P. (1992) 'Industrial Agriculture and Environmental Regulation', *Sociologica Ruralis*, 32, 4–10.

Lowe, P. and S. Ward (eds) (1998) *British Environmental Policy and Europe* (London: Routledge).

McAllister, R. (1997) *From EC to EU* (London: Routledge).

McConnell, F. (1999) Interview with the author, London, 8 October.

McDonagh, B. (1998) *Original Sin in a Brave New World* (Dublin: Institute of European Affairs).

McQuail, P. (1994) 'Mapping the Department of the Environment', Unpublished Mimeo (London: DoE).

MAFF (2001) *Consultation on Implementing the Uncultivated Land and Semi-natural Provisions of the EIA Directive* (London: MAFF).

Major, J. (1999) *John Major: The Autobiography* (London: HarperCollins).

Maloney, W. and J. Richardson (1995) *Managing Policy Change* (Edinburgh: Edinburgh University Press).

March, J. (1981) 'Footnotes to Organisational Change', *Administrative Quarterly*, 26, 563–77.

March, J. and J. Olsen (1999) 'The Institutional Dynamics of International Political Orders', *International Organisation*, 52 (4), 943–69.

Marks, G. (1993) 'Structural Policy and Multi-level Governance in the EC' in A. Cafruny and G. Rosenthal (eds), *The State of the European Community* (London: Longman).

Marks, G. (1996) 'An Actor-Centred Approach to Multi-level Governance' in C. Jeffery (ed.), *The Regional Dimension of the EU* (London: Frank Cass).

Mazey, S. and J. Richardson (1997) 'Policy Framing: Interest Groups and the Lead Up to the 1996 IGC', *West European Politics*, 20 (3), 111–33.

Menon, A. and V. Wright (1998) 'The Paradoxes of Failure', *Public Policy and Administration*, 13 (4), 46–66.

Metcalf, L. (1994) 'International Policy Co-ordination and Public Management Reform', *International Review of Administrative Sciences*, 60, 271–90.

Moore, B. (1954) 'Sewage Contamination of Coastal Bathing Waters', *Bulletin of Hygiene*, 29 (7), 689–703.

Moore, B. (1975) 'The Case Against Microbial Standards For Bathing Beaches' in A. Gameson (ed.), *Discharge of Sewage from Sea Outfalls* (Oxford: Pergamon).

Moore, B. (1977) 'The EEC Bathing Water Directive', *Marine Pollution Bulletin*, 8 (12), 269–72.

Moravcsik, A. (1991) 'Negotiating the Single European Act', *International Organisation*, 45 (1), 19–56.

Moravcsik, A. (1993) 'Preferences and Power in the EC', *Journal of Common Market Studies*, 31, 473–524.

Moravcsik, A. (1994) *Why the EC Strengthens the State*, Centre for European Studies Working Paper 52 (Cambridge, MA: Department of Government, University of Harvard).

Moravcsik, A. (1998) *The Choice For Europe* (Ithaca, NY: Cornell University Press).

Moravcsik, A. (2000) 'Integration Theory' in D. Dinan (ed.), *Encyclopaedia of the European Union* (updated edition) (London: Macmillan – now Palgrave Macmillan).

Moravcsik, A. and K. Nicolaidis (1998) 'Federal Ideals and Constitutional Realities in the Treaty of Amsterdam', *Journal of Common Market Studies*, 36 (Annual Review), 13–38.

Moravcsik, A. and K. Nicolaidis (1999) 'Explaining the Treaty of Amsterdam', *Journal of Common Market Studies*, 37, 59–84.

Nolkaemper, A. (1997) 'Habitat Protection in EC Law', *Journal of Environmental Law*, 9 (2), 271–86.

Nugent, N. (1997) 'Preparing, Waiting – and Hoping', *Journal of Common Market Studies*, 35 (Annual Review), 1–10.

OECD (1991) *Integrated Pollution Prevention and Control*, Monograph 37, April (Paris: OECD).

O'Riordan, T. and A. Weale (1989) 'Administrative Reorganisation and Policy Change', *Public Administration*, 67, 277–94.

Osborn, D. (1992) 'The Impact of EC Environmental Policies on UK Public Administration', *Environmental Policy and Administration*, 2, 199–209.

Osborn, D. (1997) 'Some Reflections on UK Environment Policy, 1970–1995', *Journal of Environmental Law*, 9 (1), 3–22.

Osborn, D. (2000) Interview with the author, London, 30 May.

Pehle, H. (1998) *Das Bundersministerium für Umwelt (BMU): Ausgegrenzt Stat Integgriert?* (Wiesbaden: Deutscher Universitats-Verlag).

Peters, B. G. (1992) 'Bureaucratic Politics and the Institutions of the EC' in A. Sbragia (ed.), *Euro-Politics* (Washington: The Brookings Institute).

Peters, B. G. (1999) *Institutional Theory in Political Science* (London: Continuum).

Peterson, J. (1995) 'Decision Making in the European Union', *European Journal of Public Policy*, 2 (1), 69–94.

Peterson, J. (1997) 'States, Societies and the EU', *West European Politics*, 20 (4), 1–23.

Peterson, J. (1999) 'The Santer Era: The European Commission in Normative, Historical and Theoretical Perspective', *Journal of European Public Policy*, 6 (1), 46–65.

Peterson, J. and E. Bomberg (1999) *Decision-Making in the EU* (London: Macmillan – now Palgrave Macmillan).

Petts, J. and P. Hills (1982) *Environmental Assessment in the UK* (Nottingham: University of Nottingham, Institute of Planning Studies).

Pierson, P. (1993) 'When Effect Becomes Cause', *World Politics*, 45 (4), 598–628.

Pierson, P. (1996) 'The Path to European Integration', *Comparative Politics*, 29 (2), 123–63.

Pierson, P. (2000a) 'Increasing Returns, Path Dependence and the Study of Politics', *American Political Science Review*, 94 (2), 251–67.

Pierson, P. (2000b) 'The Limits of Design: Explaining Institutional Origins and Change', *Governance*, 13 (4), 475–99.

Plowman, J. (2000) Interview with the author, London, 16 February.

Pollack, M. (1996) 'The New Institutionalism and EC Governance', *Governance*, 9 (4), 429–58.

Ponting, C. (1986) *Whitehall: Tragedy and Farce* (London: Hamish Hamilton).

Pootschi, B. (1998) 'The 1997 Treaty of Amsterdam: Implications for EU Environmental Law and Policy Making', *Review of International and European Community Environmental Law (RECIEL)*, 7 (1), 76–84.

Pritchard, D. (1985) 'Britain's Implementation of the EEC Birds Directive', Paper presented at a conference on 'International Wildlife Treaties', British Association of Nature Conservationists, Cambridge, September 1985.

Pritchard, D. (2000) Interview with J. Fairbrass, Sandy (Beds), 1 June.

Pryce, R. (1994) 'The Treaty Negotiations' in A. Duff, J. Pinder and R. Pryce (eds), *Maastricht and Beyond* (London: Routledge).

Purdue, M. (1997) 'The Impact of EC Environmental Law on Planning Law in the UK' in J. Holder (ed.), *The Impact of EC Environmental Law in the UK* (Chichester: John Wiley).

Radaelli, C. (2000) 'The Europeanisation of Public Policy', *European Integration Online Papers*, 4 (8) (http://eiop.or.at/eiop/texte/2000–008a.htm).

Radcliffe J. (1985) 'The Role of Politicians and Administrators in Departmental Re-organisation', *Public Administration*, 63, 201–18.

Radcliffe, J. (1991) *The Re-organisation of British Central Government* (Dartmouth: Aldershot).

Rat von Sacverstandigen für Umweltfragen (1998) *Umweltgutachten 1998: Umweltschutz Erreichtes Sichern-Neue Wege Gehen* (Stuttgart: Metzler-Poeschel).

RCEP (Royal Commission on Environmental Pollution) (1976) *Air Pollution Control: Fifth Report* (London: HMSO).

RCEP (Royal Commission on Environmental Pollution) (1984) *Tackling Pollution: Tenth Report* (London: HMSO).

Rehbinder, E. and R. Stewart (1985) *Environmental Protection Policy* (Berlin: Walter de Gruyter).

Reid, C. (1997) 'Nature Conservation Law' in J. Holder (ed.), *The Impact of EC Environmental Law in the United Kingdom* (Chichester: John Wiley).

Reynolds, F. (1998) 'Environmental Planning' in P. Lowe and S. Ward (eds), *British Environmental Policy and Europe* (London: Routledge).

Rhodes, R. A. W. (1995) 'From Prime Ministerial Power to Core Executive' in R. A. W. Rhodes and P. Dunleavy (eds), *Prime Minister, Cabinet and Core Executive* (London: Macmillan – now Palgrave Macmillan).

Rhodes, R. A. W. (1997) *Understanding Governance* (Milton Keynes: Open University Press).

Risse, T., M. Green Cowles and J. Caporaso (2000) 'Europeanization and Domestic Change: Introduction' in T. Risse *et al.* (eds), *Transforming Europe: Europeanisation and Domestic Change* (Ithaca, NY: Cornell University Press).

Risse-Kappen, T. (1996) 'Exploring the Nature of the Beast: International Relations Theory and Comparative Policy Analysis Meet the EU', *Journal of Common Market Studies*, 34 (1), 53–80.

Rosamond, B. (2000) *Theories of European Integration* (London: Macmillan – now Palgrave Macmillan).

Rose, C. (1990) *The Dirty Man of Europe* (London: Simon & Schuster).

Ross, G. (1995) *Jacques Delors and European Integration* (Cambridge: Polity Press).

Rowcliffe, J. (1999) Interview with the author, London, 9 December.

Sandholtz, W. (1993) 'Choosing Union', *International Organisation*, 47 (1), 1–40.

Sandholtz, W. (1996) 'Membership Matters: Limits of the Functional Approach to European Institutions', *Journal of Common Market Studies*, 34 (3), 403–29.

Sbragia, A. (1996) 'Environment Policy' in H. Wallace and W. Wallace (eds), *Policy Making in the EU* (Oxford: Oxford University Press).

Schnutenhaus, J. (1994) 'IPPC: The German Presidency's Initiative', *European Environmental Law Review*, December, 323–8.

Semple, A. (1999) Interview with the author, London, 9 December.

Sharp, R. (1998) 'Responding to Europeanisation' in P. Lowe and S. Ward (eds), *British Environmental Policy and Europe* (London: Routledge).

Sharp, R. (1999) Interview with the author, London, 21 December.

Shaw, R. (2000) Interview with the author, Kingston-upon-Thames, 6 March.

Sheate, W. (1997a) 'From EIA to SEA: Sustainability and Decision Making' in J. Holder (ed.), *The Impact of EC Environmental Law in the UK* (Chichester: John Wiley).

Sheate, W. (1997b) 'The EIA Amendment Directive 97/11/EC', *European Environmental Law Review*, 6 (8–9), 235–43.

Sheate, W. and R. Macrory (1989) 'Agriculture and the EC EIA Directive', *Journal of Common Market Studies*, 28, 68–81.

Skea, J. and A. Smith (1998) 'Integrating Pollution Control' in P. Lowe and S. Ward (eds), *British Environmental Policy and Europe* (London: Routledge).

Skocpol, T. (1992) *Protecting Soldiers and Mothers* (Cambridge, MA: Belknap Press).

Smith, J. (2001) 'Cultural Aspects of Europeanisation', *Public Administration*, 79, 147–66.

Smith, M. (1993) *Pressure, Power and Policy* (London: Harvester Wheatsheaf).

Smith, M. (1999) *The Core Executive in Britain* (London: Macmillan – now Palgrave Macmillan).

Smith, M., D. Richards, and D. Marsh *et al.* (2000) 'The Changing Role of Central Government Departments' in R. Rhodes (ed.), *Transforming British Government*, Volume 2 (London: Macmillan – now Palgrave Macmillan).

Smith, M., D. Marsh and D. Richards (1993) 'Central Government Departments and the Policy Process', *Public Administration*, 567–94.

Soames C. (1972) 'Whitehall into Europe', *Public Administration*, 50, 271–90.

Spence, D. (1993) 'The Role of National Civil Service in European Lobbying' in S. Mazey and J. Richardson (eds), *Lobbying in the European Community* (Oxford: Oxford University Press).

Stack, F. (1983) 'The Imperatives of Participation' in F. Gregory (ed.), *Dilemmas of Government* (Oxford: Martin Robertson).

Stetter, S. (2001) 'Greening the Treaty: Maastricht, Amsterdam and Nice', *European Environmental Law Review*, May, 150–9.

Summerton, N. (1999) Interview with the author, Oxford, 8 December.

Taylor, P. (1989) 'The New Dynamics of EC Integration' in J. Lodge (ed.), *The EC and the Challenge of the Future* (London: Pinter).

Thairs, E. (1998) 'Business Lobbying on the Environment' in P. Lowe and S. Ward (eds), *British Environmental Policy and Europe* (London: Routledge).

Thatcher, M. (1993) *The Downing Street Years* (London: HarperCollins).

Thelen, K. (1999) 'Historical Institutionalism in Comparative Politics', *Annual Review of Political Science*, 2, 369–404.

Thelen, K. and S. Steinmo (1992) 'Historical Institutionalism in Comparative Politics' in S. Steinmo K. Thelen and F. Longstreth (eds), *Structuring Politics* (Cambridge: Cambridge University Press).

Therivel, R., E. Wilson, S. Thompson, D. Heaney and D. Pritchard (1992) *Strategic Environmental Assessment* (London: Earthscan).

Tinker, J. (1975) 'Pollution in Brussels: Logic or Lunacy?', *New Scientist*, 68, 970.

Toonen, T. (1992) 'Europe of the Administrations', *Public Administration Review*, 52, 108–15.

Trippier, D. (1991) 'Development in Environmental Policy at the European Level', *European Environment*, 1 (6), 7–10.

Vandermeersch, D. (1987) 'The SEA and the Environmental Policy of the EEC', *European Law Review*, 12, 407–29.

von Moltke, K. (1983) 'Influences on EC Environmental Policy' in R. Macrory (ed.), *Britain, Europe and the Environment* (London: ICCET, Imperial College London).

von Moltke, K. and N. Haigh (1981) 'Major Issues for 1981', *Environmental Policy and Law*, 7, 23–31.

Waldegrave, W. (1985) 'The British Approach', *Environmental Policy and Law*, 15 (3–4), 106–15.

Wallace, H. (1995) 'Britain out on a Limb?', *Political Quarterly*, 66 (1), 46–58.

Wallace, H. (1996) 'Relations Between the EU and the British Administration' in Y. Mény P. Muller and J.-L. Quermonne (eds), *Adjusting to Europe* (London: Routledge).

Wallace, H. (1997) 'At Odds With Europe', *Political Studies*, XLV, 677–88.

Wallace, H. (1999) 'Review Section Symposium: The Choice For Europe', *Journal of European Public Policy*, 6 (1), 155–9.

Wallace, H. (2000a) 'Europeanisation and Globalisation', *New Political Economy*, 5 (3), 369–82.

Wallace, H. (2000b) 'The Domestication of Europe and the Limits to Globalisation', Paper for the International Political Studies Association (IPSA) World Congress, August 2000 (University of Sussex: Sussex European Institute).

Wallace, H. and W. Wallace (1973) 'The Impact of Community Membership on the British Machinery of Government', *Journal of Common Market Studies*, 11 (4), 243–62.

Wallace, H. and W. Wallace (eds) (2000) *Policy Making in the EU* (Oxford: Oxford University Press).

Ward, H. (1993) 'Purity and Danger' in M. Mills (ed.), *Prevention, Health and British Politics* (Aldershot: Avebury).

Ward, S. (1997) 'The IGC and the Current State of EU Environmental Policy: Consolidation or Roll Back?', *Environmental Politics*, 6 (1), 178–84.

Water Services Association (1996) *Water: The Facts* (London: Water Services Association).

Wathern, P. (1988) 'The EIA Directive of the EC' in P. Wathern (ed.), *EIA: Theory and Practice* (London: Routledge).

Wathern, P. (1989) 'Implementing Supranational Policy: EIA in the UK' in P. Bartlett (ed.), *Policy Through International Affairs* (London: Greenwood Press).

Weale, A. (1992) *The New Politics of Pollution* (Manchester: Manchester University Press).

Weale, A. (1996) 'Environmental Rules and Rule-Making in the EU', *Journal of European Public Policy*, 3, 594–611.

Weale, A. (1997) 'The United Kingdom' in M. Janicke and H. Weidner (eds), *National Environmental Policies* (Berlin: Springer).

Weale, A. *et al.* (2000) *Environmental Governance in Europe* (Oxford: Oxford University Press).

Weale, A. and A. Williams (1992) 'Between Economy and Ecology: The Single Market and the Integration of Environmental Policy', *Environmental Politics*, 1 (4), 45–64.

Weiler, J. (1991) 'The Transformation of Europe', *The Yale Law Journal*, 100, 2403–83.

Wilkinson, D. (1990) *Greening the Treaty: Strengthening Environmental Policy in the Treaty of Rome* (London: IEEP).

Wilkinson, D. (1992) 'Maastricht and the Environment', *Journal of Environmental Law*, 4 (2), 221–39.

Wilkinson, D., K. Bishop and A. Tewdwr-Jones (1998) *The Impact of the EU on the UK Planning System* (London: DETR).

Williams, R. (1986) 'EC Environment Policy, Land Use Planning and Pollution Control', *Policy and Politics*, 14 (1), 93–106.

Williams, R. (1991) 'Placing Britain in Europe: Four Issues in Spatial Planning', *Town Planning Review*, 62 (3), 331–40.

Willis, V. (1982) *Britons in Brussels* (London: Policy Studies Institute).

Wils, J. (1994) 'The Birds Directive 15 Years Later', *Journal of Environmental Law*, 6 (2), 219–42.

Wincott, D. (1995) 'Institutional Interaction and European Integration', *Journal of Common Market Studies*, 33 (4), 597–609.

Winter, M. (1996) *Rural Politics* (London: Routledge).

Wood, C. (1995) *Environmental Impact Assessment* (Harlow: Longman).

Wood, C. and C. Jones (1991) *Monitoring Environmental Assessment and Planning* (London: HMSO).

Wood, C. and G. McDonic (1988) 'Environmental Assessment: Challenge and Opportunity', *The Planner*, 7 July, 12–18.

Wurzel, R. (1993) 'Environmental Policy' in J. Lodge (ed.), *The European Community and the Challenge of the Future* (London: Pinter).

Wurzel, R. (1999) 'Britain, Germany and the European Union', Unpublished PhD thesis (London: LSE).

Wurzel, R. (2002) *Environmental Policy-Making in Britain, Germany and the European Union* (Manchester: Manchester University Press).

WWF (2001) *A Race to Protect Europe's Natural Heritage* (WWF-European Policy Office: Brussels), http://www.panda.org/publications/sspub.cfm

WWF, FoE and EEB (1990) *Greening the Treaty*, November (Brussels: EEB).

Young, H. (1998) *This Blessed Plot* (London: Macmillan – now Palgrave).

Young, H. and A. Sloman (1982) *No Minister* (London: BBC).

Young, S. C. (1995) 'Running Up the Down Escalator' in T. Gray (ed.), *UK Environmental Policy in the 1990s* (London: Macmillan – now Palgrave Macmillan).

Zito, A. (2002) 'Task Expansion' in A. Jordan (ed.), *Environmental Policy in the European Union* (London: Earthscan).

Index